THE
RIPENING SUN

Patricia Atkinson

arrow books

Published in the United Kingdom in 2004 by Arrow Books

1 3 5 7 9 10 8 6 4 2

First published in the United Kingdom in 2003 by Century

Arrow Books
The Random House Group Limited
20 Vauxhall Bridge Road, London, SW1V 2SA

Random House Australia (Pty) Limited
20 Alfred Street, Milsons Point, Sydney,
New South Wales 2061, Australia

Random House New Zealand Limited
18 Poland Road, Glenfield
Auckland 10, New Zealand

Random House (Pty) Limited
Endulini, 5a Jubilee Road, Parktown 2193, South Africa

The Random House Group Limited Reg. No. 954009

www.randomhouse.co.uk

A CIP catalogue record for this book
is available from the British Library

Papers used by Random House are natural, recyclable products made from wood
grown in sustainable forests. The manufacturing processes conform to the
environmental regulations of the country of origin

ISBN 0 09 947154 X

Printed and bound in Great Britain by
Bookmarque Ltd, Croydon, Surrey

To Amy, Beth
and the Man on the Edge

Acknowledgements

I would like to thank the people who appear in this book and all those who don't, — too numerous to mention — who have helped me and without whom I could never have succeeded. The fact that I can relate this story I owe to James, who brought me here, and the fact that it is written at all I owe to Oliver Johnson for commissioning the book and for offering his skill, patience and generosity in the editing of it. My thanks also to Jilly Farrer Brown for her line drawings, Anne Brewin for her photographs, and most especially to Nigel Farrow, whose advice and encouragement were invaluable.

The château and vineyards of Gageac, looking north

KEY TO MAP – *The Ripening Sun*

1. The château at Gageac et Rouillac
2. Yves and Juliana's house
3. Geoffroy's house
4. Jean de la Verrie's *chai*
5. Clos d'Yvigne
6. The *monument du mort*
7. The church
8. The cemetery

9. The iron cross
10. The old bakery
11. Roger and Pepita's house
12. Madame Cholet's house
13. Gilles' house
14. The dilapidated cross
15. The statue of Our Lady
16. Eida

Autumn

I solitary court
The inspiring breeze, and meditate the book
Of Nature, ever open, aiming thence
Warm from the heart to learn the moral song
And, as I steal along the sunny wall,
Where autumn basks, with fruit empurpled deep,
My pleasing theme continual prompts my thought –

Presents the downy peach, the shining plum
With a fine bluish mist of animals
Clouded, the ruddy nectarine, and dark
Beneath his ample leaf the luscious fig,
The vine too here her curling tendrils shoots,
Hangs out her clusters glowing to the south,
And scarcely wishes for a warmer sky.

James Thomson 1700–1748
(Autumn 669–82)

Prologue — August 2002

AS I LOOK TOWARDS CLOS D'YVIGNE FROM THE CHURCH OF GAGEAC, the morning sun casts dappled light onto the new wall that seems as though it has always been there. It bears the name of Clos d'Yvigne in relief, red lettering against sand and white stone. Behind the wall is a small courtyard with tamarisk trees and wisteria. A recently constructed tasting room, with adjoining office and stocking depot, is to the right. To the left of the courtyard is my house and behind it the winery, the working heart of the vineyard. It contains wine in vats and barrels, along with all the tools of winemaking; pumps, vats, pipes and presses.

In the tiny orchard of fruit trees beyond the winery sits a Massey Fergusson tractor. Further on still are the vines, a *parcelle* of Merlot in the foreground, their grapes just beginning to change colour from green to deep bishop's purple. I have twenty-one hectares of them, set on the beautiful ridge of hills that run along the left side of the Dordogne in south-west France.

The large, dense leaves clothing the vines reach up to the sun, soaking up nourishment and transferring it to the grapes. I calculate that it will probably be October before I harvest them. Today is Sunday, an afternoon in mid-August. It's the day of the week that I give to myself, in itself a form of liberation. If I want to work I can, but it's officially a day of rest.

I walk up through the *parcelles* of vines towards a dilapidated cross which sits next to a derelict house. In front of the house is a tree, under which is the old wooden seat where Madame Cholet used to sit in the afternoons.

I am standing on the highest point of my land, the plateau a vast, shallow bowl covered with vines, its perimeters studded by historical monuments. To the west is the fortress-like silhouette of the chateau of Saussignac, the tall spire of the church close beside it. To the south is my house, and beyond it the beautiful chateau of Gageac, home to the de la Verrie family. To the north, the land drops away steeply to the valley of the Dordogne. Far below, the fields and vines create a tapestry of yellow and green.

Deep in the river valley in front of me lies the historic town of Bergerac, fought over by Protestants and Catholics in the Wars of Religion and by England and France in the Hundred Years War. To the right and in the foreground is a copse of trees, behind which lies Pomport where the first battle of that war was fought, and interspersed amongst the villages are fruit trees and vines that have been cultivated here for hundreds of years.

I gaze at the horizon and the land, timeless, yet changing; defined not only by history and culture, but also by an interaction with the land, the vines and nature. The confines of my life for the last twelve years have been these landscapes, both near and far. Having come here precipitately, I put down roots and learned to live off this land and these vines.

It is hot. I look at the vines in the foreground; graceful, contoured carpets of lush green, sweeping out in front and to the right of me. Their gentle, sinuous curves follow the contours of the landscape, swathes of form and colour. A summer haze hangs over them like a mantle.

The drone of a distant tractor lends a communality of spirit. I stop for a moment. Hands cupped over my eyes, adjusting to the shimmering glare of the sun, I look up towards the cemetery with its dark, mysterious cypress trees and onwards, to the vines up on the ridge. Jean de la Verrie is slowly ploughing, up and down his vines, the rhythmic sound of the motor accentuating the heat.

Chapter 1

A New Life – May 1990

'DELICIOUS ROSÉ. GOD, I NEED THIS DRINK.'

Laura has taken off her shoes and stretched herself out on the chair in the courtyard. It's the last day of May, a hot, balmy evening with the scent of the privet hedge overpowering. The singing of the cicadas creates an ambience of mediterranean summers and the sun-drenched courtyard gives up its luxurious heat in waves.

I am sitting opposite my husband, James, and two friends, Charles and Laura, who have come over to help us move. The sun is setting, deep magenta as it slips away behind and to the right of the château of Gageac, towers and steeply gabled roof silhouetted in its last rays. 'It's exquisite,' I murmur, watching as the sky transforms itself, suffused first with soft light and colour, then with ripples of gold, orange and pink; a sea of changing colours, casting its spell over the château and me.

Charles and James have bought a selection of crudités from the local market, along with some bottles of wine for supper. The courtyard we are sitting in is actually a wilderness of bald grass and gravel, with a lime tree in one corner and a makeshift shower, surrounded by the untended privet hedge. I feel vaguely disorientated but happy, luxuriating in the warmth of the evening as

5

it washes over me. I am here in France on the last day of May 1990 and about to embark on a new life.

Laura and I have just arrived, having been on an unintended excursion around France. Overbooked on our scheduled flight to Bordeaux from Heathrow, British Airways offered us an alternative flight to Bergerac, via Lyon. There was a strike at Lyon. Eleven hours later we arrived at Gageac.

We chatter together, laughing and glad to be here. I look at James, whose delight is evident. 'We'll show you the house in a minute,' he says, smiling and looking at Charles – who looks at me, then Laura, with eyes wide and lips pursed. 'Yes,' he mutters.

I have seen the house only once before, a year ago on a weekend trip to France after James bought it. It is not my dream home. Situated on a road in a small hamlet, it has no view and no privacy. When I last saw it, the outside was plastered with what looked like white cowpats of cement, creatively sculpted by some builder, or perhaps even the owner. You stepped from the road into the kitchen, which was painted dark brown and decorated with yellow, flowered wallpaper. It was filled with fridges, boxes of melons, tomatoes and a large French country sideboard. The owners were sitting inside and welcomed me. As I didn't speak French and they didn't speak English, there was not a lot to say. I felt ill at ease, an intruder in their home. They didn't offer to show me any more of the house and, truth to tell, I was glad to escape. And now I am back, to live in it.

'At least the lorries arrived before you,' Charles is saying to us. 'A close thing. We didn't think they would.'

The two lorries carrying our possessions set off from England the day before Charles and James. Arrangements were made for them to meet each other at Cherbourg and travel on together, but somehow they didn't. Laura and I laugh as the men recount their journey to Périgueux to meet the customs & excise people, the lorries and the drivers. The bureaucracy, the papers, and dealings with the officials were fraught with misunderstandings. We knew something about the paperwork that might be involved from our purchase of the house. A French friend, Rosemarie Doughty, married to an Englishman and living in the area, had helped us through the minefield of bureaucracy, increased due to the purchase of four hectares of vines that came with the house. One of her sons, Richard, has a vineyard nearby and James made a financial arrangement with him to look after the vines before our arrival and help us in our first year.

'Come on,' says Laura, rising from the chair with her glass, 'Let's have a look at it before we eat.' Glasses in hands, we step into the semi-darkness of the house, a dramatic contrast to the bright light of the courtyard. In the next few minutes I see, for the first time, the house in its entirety.

The builders haven't progressed as quickly as we envisaged. There is no kitchen to speak of, only the elements of one in the form of the MFI kit bought months before in England. Instead of a window, a gaping hole in one of the walls looks out onto a wilderness which is the garden. We wander into the *salon*, where a mountain of furniture and belongings that arrived earlier in the day is sitting on a newly laid concrete floor. There is no electricity.

A wooden stepladder gives access to the first floor, as there is also no staircase. Upstairs, the bedrooms are empty and dusty. There are beds, but they are hidden among the mountain of furniture downstairs. Although there is no bath, the absence of one is not too much of a problem as there is no bathroom either – only a lavatory, installed just today.

The chaos of it is daunting. None of it, however, can dampen my spirits, such is my delight in being here at last. Laura and I laugh as we clamber over the mountain of furniture to reach the courtyard again, our eyes dazzled by the bright sun and warmth outside.

'Mmm,' says Laura. 'Let's have supper.'

'Here's to a new life,' says Charles, lifting his glass as we rein-stall ourselves in the courtyard. '*La vie est belle,*' he adds, and we all laugh. According to Laura, it's the only French Charles knows.

'God, it's good to be here,' I say.

'It's good to be eating something!' says Laura.

The rosé is delicious, and the crudités too; tiny delicacies of meats, vegetables in unusual sauces and flavoured with herbs and spices, along with fish dishes, pâtés and cheese. The bread is soft with a crunchy crust and cheese has never tasted better.

A deep ink blue with clear, sparkling stars has replaced the spectrum of changing colours in the sky. The cicadas are still call-ing, but more softly. Sighs from under the privet signal that Sam and Luke, two red setter puppies, are asleep there and dreaming. Laura looks relaxed as she presses the cold glass of wine to her cheek. 'Mmm' she murmurs softly. More and more stars appear, dramatic against the black night sky.

We bask in the euphoria of the evening, intoxicated by the wine, the perfume of the privet, the beauty of the sunset and the pleas-ure of each other's company. 'The plumber's going to try and have water in the house by tomorrow or the day after,' says James. 'He's the electrician too. He and his son have been here all day.'

I am half-listening, my mind wandering in a desultory, dream-like state and reflecting back to the evening over a year ago when we decided to move here. James wanted to change his life and do something different. The idyll of France with its culture and lifestyle had long been one of his dreams.

He started looking for a property in France. After a search and with the help of our friend Rosemarie Doughty, we found a house that was something close to my imaginings: a lovely stone build-ing; a wonderful view; a river nearby and privacy. We made an offer which was accepted, then returned home to England. The day before James was due to come back to France to sign the contract, the seller told us he had changed his mind. James flew to France anyway. He rang that evening. He'd found something else. 'You'll like it, darling,' he said. 'The house can be anything you want it to be. And there are vines.'

And now I am sitting in a courtyard in France, on a hot, balmy evening. James's dream is a reality. I look at his face as he talks to Charles. It's full of hope for the future, full of happiness. 'We've set up a makeshift shower over there,' James is saying to me now, pointing to a corner of the courtyard. Apart from the lavatory, there is no running water in the house. He is flushed with the sun, the wine and delight in being here. 'We're going to start on the kitchen cabinets tomorrow.'

'I'm being bitten by mosquitoes,' announces Laura, 'and it's time we went to bed.'

'Keep moving and they won't get you,' says Charles, and we leave the candlelit supper and go in search of mattresses. Somehow we manage to transport them up the old wooden ladder, and make up beds on the floor of the two bedrooms. James suggests a quick stroll before bed and we leave Charles and Laura to the shower and wander through the vines.

With the gentle but insistent call of the cicadas, we walk in a fairy landscape. Paths of moonlight with glow-worms light our way among the rows of vines. Moon shadows cast a strange and magical light on the terrain and we feel transported to entirely another place.

A hot day – the first day of June. It is 7 o'clock in the morning. Although the window panes in the bedroom are thick with dust, the sun is streaming in through them. I have spent my first night of a new life in France. The dust on the bedroom floor made it almost impossible to breathe in the night, and we are surrounded by mountains of black bin bags bursting with clothes. Now I remember: there is no bathroom and there are no stairs, only the ladder. There are also no cupboards. But a great sense of adventure invades me with a positive delight to be here at last.

I fling open the window and look out of it to see the bay horse in the field opposite looking back at me. She is standing by the hedge, opposite the kitchen door – almost in the kitchen, it seems to me. She has a quizzical look on her face, which I take to be a welcome of sorts. Beyond her field is a long, low stone building and beyond that are vines. Climbing a gentle slope to a wooded ridge, they look green, fresh and beautiful. To the left of them in the distance is the cemetery, thick with tall cypress trees. There is a large cross at the top of the road that leads down from the cemetery to our house and to the church opposite. It is a simple country church, small with a plain classical facade. Above the door is a lintel, which carries a stone cross in its centre, replicated again at the apex of the triangular façade. Below the apex is a small open arch, in which is suspended a lone bell.

There is a small grotto at the side of the church occupied by a statue of the Virgin and another statue of her across the horse's field. It is life size and is made of a silvery metal that glints in the sunlight. The silver Madonna, I decide to call her.

'*Madame, regardez! C'est vos chiens!*' My daydream is broken by a small, dark weather-beaten woman. She is standing between the horse and the house, holding a bundle wrapped loosely in newspaper. There follows much animated discourse. She is wearing an apron over a lacy petticoat, which falls below the hemline of her overall. A woollen hat is perched on her head, covering one eye and she wears ankle socks, slippers and *sabots*.

'James, quick, get up, someone's speaking French to me!' I shout behind me. James leaps out of bed while I smile stupidly at her and wave. So begins the first of many one-sided discussions with Pepita, a Spanish neighbour, and so also begins the infamy of Sam and Luke, a scourge on the neighbourhood for the first year of our life here.

A few moments later and we are now all at the kitchen door

– Charles and Laura, James, who has hurriedly dressed, and me. '*Bonjour, Madame*,' we say sheepishly.

She shows us the bundle in the newspaper. To my horror it contains two dead chickens.

James is speaking to her in French. I squirm in shame, not only because of the dogs and the dead chickens, but because I am unable to respond in any way other than mute guilt, which I can convey only in much eyeball rolling and sighs of regret. I make an instant decision to learn French as quickly as possible. In the meantime, there appears to be no way to end the discussion, as James's words are falling on deaf ears.

She continues, in full flow, with much flinging of arms and pointing at the chickens, occasionally reaching to her hat and repositioning it. I stand behind James, and Charles and Laura stand behind me. James offers her money, which calms the waters momentarily. The flow of incomprehensible words falters and James grabs the opportunity to wish her *bonne journée* and close the door. The horse continues to stare at me quizzically through the panes of glass in the door. The dogs, who have been standing beside us listening intently, wag their tails and nuzzle my hand as we all regard each other in silence.

'I've seen where she lives. It's just up the road,' says Charles. 'She's got a lot of flowers. Buy her a plant, she'll be fine.'

Laura looks at the dogs, then at me and says: 'If I were you, I should buy a dozen or so and leave them outside the kitchen door with a note saying 'Exchange your dead chickens for plants here'.

'This house is a dreaming house,' says Laura over breakfast. 'I had vivid dreams last night. Did you?'

My dreams had been broken by constant bouts of breathless-ness from the dust. Not to mention the scuttling of nameless wildlife under the floor and in the ceiling. 'First job is to wash down the floors in the bedrooms and put up the beds,' I say.

11

Which is what we do, with cold water from the makeshift shower in the garden and much elbow grease. The men go in search of bed legs and frames in the lorry load of furniture in the *salon*, and I go off to find a plant for Pepita.

As I walk up the road to the house of the dead chickens, plant in hand, I rehearse to myself: '*Bonjour, je m'appelle Madame Atkinson, je suis desolée pour les poulets.*'

Pepita's house is a small, low building with a panoply of flow-ers in front of it. Roses of all descriptions mingle with carnations and large yellow and orange chrysanthemums. In front of and behind the house are old rusty buckets filled with seedlings, and empty tins for feeding the chickens, geese, turkeys, ducks and rabbits that live in the clusters of small outbuildings behind the house, giving the place a bedraggled but romantic air.

I rehearse my speech one last time and knock at the door. From within I hear chair legs being scraped back, followed by mumblings, then silence. I knock again and hear someone shout-ing behind the door. The door opens and before me stands a vision that eradicates all memory of the speech I had rehearsed – a nose, a huge, enormous cauliflower nose. I am mesmerised by it. It is truly enormous, wide and mountainous. The face behind it is a crater, with the nose a series of them, each one enveloped by another. It stands in the middle of the face and spreads itself over it, glorious in its supremacy. An equally enormous stomach accompanies it.

The possessor of the nose and stomach is as taken aback by me as I am by him. I have obviously disturbed him. He hasn't completely woken up, or, for that matter, dressed. He has failed in an attempt to pull his blue strapped worker's overall over his astonishing stomach.

He says something to me, but I understand nothing and have the same feeling of shame and inadequacy I had earlier with Pepita.

I am unable to explain about the dogs, the chickens, or the plant for his wife, so mutter a *'bonjour'*, interspersed with a *'poulet'* and a *'merçi beaucoup'* and retreat down the road to safety. This is my first meeting with Roger Merlos.

The morning is hot and beautiful. As I walk back down the road from Roger and Pepita's I see a tractor turning into the vineyard next to our house. It's only my second day in France and already I feel immersed in French life and culture. I know I'm about to immerse myself even deeper into it as a man descends from the tractor. He is our worker, Boujema, found for us by Richard Doughty.

Well built, stocky, with a tanned, lined face and black, intelligent eyes, he strides towards me, laughing and gesticulating. He holds out his hand, and shouts *'Bonjour!'* again over the noise of the tractor, which is still running. I say *bonjour* sheepishly and shake his hand. His hands are enormous, his handshake strong and his smile wide. He talks and laughs simultaneously as he introduces himself. He points towards the spraying machine attached to the tractor, then at the washing we had hung on the line last night, then he leads me to the vines and points to the grapes, laughing and gesticulating.

I look at the bunches of grapes on the vine for the first time at close quarters. They are only three inches or so in length, tiny green hard balls. In fact they look rather like a diagram of a molecule. In contrast, the leaves seem enormous, green veined with serrated edges, scores to each vine. Boujema points again to the washing and the car, remounts the tractor, revs it up to full throttle, waves and sets off. The air is filled with a fine blue spray that covers everything. Now I understand; washing in, car moved, door closed.

'Ah, bonjour Madame.' I meet the plumber, the electrician and the carpenter. The house is teeming with people. Charles, James and

Laura are finishing breakfast, while the plumber drills holes in the newly laid cement floor in the *salon*, carefully positioning himself and his tools amidst the furniture mountain. The carpenter is measuring a corner of the room for the proposed staircase and the plumber's son is busily unravelling metres of electric wire. 'Okay,' says Charles. 'Let's get down to it.' Charles and James are going to fit the kitchen cupboards. Laura and I gather up paint pots and brushes and ascend the ladder to start painting the upstairs rooms.

We have got used to living with builders. Serge, the mason, is hard at work with his two employees, replacing tiles on the roof. In his forties, he is tall and tanned with a slow walk and an intelligent face. He works methodically, descending the ladder periodically to gather up more tiles. Each time he descends he lights a cigarette, then climbs back up, cigarette in mouth, carrying rows of tiles in the hod on his right shoulder. He wears only shorts, the sun beating down on his torso, which is already a deep, rich brown. His two workers are pointing the outside walls that have been denuded of the white cowpats. They, too, are tanned dark brown and I wonder how their skins can withstand such intense heat.

The cement floor in the *salon* has dried and the furniture mountain, although still there, is diminishing slowly. The kitchen cabinets are also installed and operational, thanks to Charles and James. The walls of the kitchen, which were covered in some strange form of carpeting as well as the flowered paper have had a coat of paint and are now white. The carpeting, we discovered, was there to hide the saltpetre and damp in the wall. Laura and I have painted room after room, placing appropriate baggage in them. Endless washing of floors and general cleaning have created some sense of order. We still have no staircase, but have become adept at scrambling up and down the rickety ladder. Most important, we now have a bathroom and lavatory upstairs.

*   *   *

Two weeks pass in earnest work and James is due to leave for England to do some consulting work. Charles's and Laura's time is up too. 'I'll only be gone a week or so,' James reassures me as he stands in front of me, a briefcase in his hand. I think back to the time a year ago when we made the decision to come out here.

With two children, a son, John who has already left home, and Chantal, who is about to take A levels, along with James's three children, Emma, Sophie and Andrew of about the same age, I hadn't considered anything other than the life we had. Spending ten days or so each spring in the Loire valley, often with Charles and Laura, was something we always looked forward to. Having a holiday home in France was a possibility, but to move there permanently? James suggested we spend a month thinking it over before making any decision.

Two weeks after James made his suggestion, the decision was made in principle. The children liked the idea of their home being in France rather than in England and I liked my ideal of France – sun, beautiful countryside, good food and wine. Not speaking French and knowing nothing of rural life here didn't disturb my idyll much then. I brushed it from my mind with a vague idea of resolving the problem later. James also wanted some vines; not many, but enough for a small, recreational business while he continued with a number of financial consultancies, grouping the work into one week or so a month spent back in England.

And now we are living the idyll. James looks incongruous, standing in front of me in a suit. 'See you in a week' he smiles and is gone.

The sound of Boujema's tractor announces his imminent arrival and I hurry out to gather in the washing. He arrives twice a week, usually on a tractor and always with strange appendages attached to the back of it. Today it has a spraying machine on the back to spray the vines. This he does every ten days or so, in spite of the fact that the leaves at the bottom of the vines are still bluish from the last one. He revs his tractor up to what sounds like maximum and sets off through the rows, spraying the blue liquid on them. Sometimes he comes to trim the leaves, at other times he drags a huge mowing machine through, cutting the grass between the vines.

I look again at the vines. In ten days, the grapes have grown considerably. They are now definitely grape sized. They are still green, a fresh, young green, with tiny blue specks. The leaves around them are blue from the spray too, but not so the leaves at the top of the vine. The tendrils that ten days ago were green, delicate shoots are now wood-like and firmly attached to the wires on the canopy. I know that we have four hectares of vines. I haven't, as yet, much of an idea of how much or what four hectares of vines represents.

It's my first evening on my own and I decide to apply myself to French study. Armed with Chantal's old school course books, Longman's *French Au Courant* level one, Harrap's *Locutions Vivante* and a large French dictionary, and determined to get to grips with the language, I open the *Au Courant* book.

First chapter, first page: '*à quoi bon apprendre le français?*' Why learn French? '*Pour pouvoir apprécier la culture et la littérature, les films et le théâtre, la radio et la télévision.*' The radio I have already tried and abandoned. Someone suggested always having it on as background to acquaint myself with the sound of French. The speed at which the commentators talk is alarming. Although I concentrate , it proves impossible to work out even what the subject is, let alone individual words or phrases. My incomprehension is total and instantly depresses me.

As for the culture, this is still a mystery, apart from my romantic and unrealistic view of it, and films and theatre haven't entered my consciousness as even a vague possibility.

I open the french windows. It is hot; very hot. A sultry evening, balmy and without air. The dogs are lying asleep on the tiled floor next to me, occasionally moving a leg or their chins for a change of place and another cool tile. Hours pass and night falls as I sit at the kitchen table trying to make sense of the first chapter and the exercises. The cicadas are louder this evening than on previous evenings, and more insistent. Bats fly around the only street light, at one corner of the horse's field opposite the house. They must be hunting small flies and mosquitoes.

There is a loud thud as something hits the floor. The dogs open an eye then leap up to inspect. I turn and look down to see an enormous black beetle with huge legs, and pincers as antennae. It advances towards me with a slow but determined gait. My heart stops momentarily before panic takes over.

I flee to the salon, cement floor, furniture mountain, stacked pictures and stuffed bags, and up the ladder to the bedroom where the window is open. Bats continue to swirl around the lamp opposite it and I see, at close quarters now, that what they are hunting are dozens of these beetles. Two of the bats fly into the room and swerve within six inches of my head. I gaze in horror for a second, then close all the windows in stomach-gnawing panic and fear.

I decide to risk going back down to the *salon*. I have to do something, even if it's just to close the window downstairs. I approach the kitchen, the dogs by my side. They have now been struck by the seriousness of the moment and advance as gingerly as I do, tails wagging gently.

On the kitchen floor, the huge bug is lying on its back, occasionally moving its front legs. If it did have malicious intent, it has no longer. I approach and suddenly feel great pity for it.

With the aid of a long handled floor brush, I gently right its body. It moves slowly. With care, I remove him from the kitchen floor to the road, spotlighted underneath the outside lamp. The dogs regard it with benevolence for a moment, then Luke pounces on it and eats it.

As I shut the windows, there is more movement in the middle distance. A procession of a hundred or so people advance from the direction of the cemetery. They carry candleholders with small opaque shields, protecting the burning candles in them against the light summer breeze and accentuating their flame. The darkness is flooded with their golden hues. A low chant accompanies them as they pass by the wrought iron cross at the bottom of the road, then head directly towards me and the church. I gaze at them, spellbound and enchanted. As they reach the church, they stop momentarily near the monument in front of it, their words now clear.

They are singing *Ave Maria* softly as they slowly pass the house. I watch in wonder, then rush upstairs to see more of them from one of the bedroom windows. They continue their progress along the road around the horse's field, reaching the statue of the Virgin at the far corner, where they gather, the flames from their candles flickering light into the dark while the soft melody floats on the air towards me. What is this procession? Is it the feast day of the Virgin? I feel suddenly a sense of history, events of the past here in Gageac merging with the present; a permanence, a tradition of village life centring in and around the château and church.

They resume their pilgrimage around the field and on to the church and its small grotto, where they kneel in prayer. Then, one by one, they extinguish their candles and drift away in small groups, leaving me baffled yet with a sense of calm.

All is silent, save for the cicadas as I drift into sleep, the pleasure of a soft bed with clean sheets a prelude to it. The bedroom has two windows, one looking out onto the horse and the street light of Gageac and the other onto the courtyard, dominated by the beautiful lime tree. My bed is next to the courtyard window and as I slip into sleep I mull over the events of the evening.

How extraordinary my first night on my own in Gageac has been. Watching the past, present and future merge in a religious rite that has always been there. I must find out more about the church of Gageac and I must visit the cemetery. How long has Gageac existed? How absurd not to know anything about it. And how absurd to have reacted so neurotically to the small black bug earlier in the evening, even if it wasn't that small. And the bats . . . I sit up in sudden fear and panic. A low, hissing noise emanates from the courtyard. I am absolutely not imagining it! I listen intently, heart pounding. There is heavy breathing, deep and low, outside the window.

A screech fills the air. I sit bolt upright in the bed, frozen by horror and gripped by panic and fear. The dogs bark downstairs and I catch briefly from the window the flight of a white owl from the lime tree.

Chapter 2

'CHANTAL, *PLEASE* COME WITH ME TO THE CHÂTEAU. I'LL CARRY THE flowers and you say thank you in French.'

My daughter has arrived from England for two weeks and James is back too. I hardly noticed the time pass after the bug experience. I missed James, but spoke to him often on the phone. I missed the children too, but was cheered by their imminent arrival. And I had worked hard on the house. We now had a staircase and electicity.

'But Mum, what if she replies?'

Madame de la Verrie, *la Comtesse du château*, called while I was out trying to grapple with French money, the French language and French drivers. She had brought some flowers to welcome us to Gageac – a kind and generous gesture which now requires another nerve-wracking attempt at conversation. Chantal, at least, knows some French. She was home when Madame de la Verrie called.

'What's she like?' I ask.

'Very nice; she just wanted to welcome us to Gageac. We didn't talk a lot.'

We set off towards the château. Approaching it from the bottom of the road, the sunlight glinting on the richly coloured tiles of the roof contrasts with the pale grey stone of the walls. The near tower and the sheer scale of the château as a whole are both awesome and beautiful. Each of the towers, one slightly smaller than the other, has two weather vanes, delicate silhouettes against the blue sky. 'Wow, it's beautiful,' murmurs Chantal. We turn left at the bottom of the road and follow the low wall alongside the grounds of the château, past a side gate with a small avenue of beech trees, eight each side, which leads to an entrance; a small arch with wooden doors, painted in pale grey. There is a notice set in the wall. It tells us that the oldest section of the château dates from the eleventh century and that it was originally a fortified prison. An enigmatic statement at the bottom of the notice informs us that 'walking around the garden will be tolerated'.

'That's not the entrance, Mum. Keep going,' says Chantal.

As we turn the corner towards the main entrance, its beauty strikes us again. The high walls of the château are surrounded by a wide moat, which is now grassed. A small roofed turret on one corner compliments the change in style on the top of the wall from castellated to tiled. The main, gated arch leads into a courtyard. We enter and are both taken aback by the stillness, the heat, and an overriding sensation of peace. It is silent. The feeling of history and permanence that struck me as the procession passed by the house returns. Humanity changes, the rest is timeless. Doves coo up in the towers, the only sound, echoing the serenity of the place.

On the wall of the château in front of us is a wisteria, dazzling pale lilac against the grey and white stone. The main doors are open, leading to a central staircase. 'God, Mum! It's gorgeous.' Chantal whispers. As we approach the doors, Madame de la Verrie appears. In her seventies, she is small, slim and elegant, with white hair pulled back into a bun. She welcomes us warmly and invites us into the main *salon* to the right of the staircase and through a large dining room.

Its simplicity and elegance are captivating. A large, pale carpet, probably from Aubusson, covers a section of the floor, which is dressed with small and beautiful terracotta tiles, deeply patined. Louis XIV chairs are scattered here and there, with a small sofa. To the side is a writing desk with photos and some roses.

Monsieur de la Verrie arrives, smaller still than his wife, wearing a jacket and hat, which he removes in order to bow and shake our hands. He is eighty-six and has spent his life here, as did his mother before him. I hand over the flowers to Madame. Their generous welcome is evident, even though I can't understand them. Chantal is thanking her for the flowers and her welcome, I am smiling stupidly, mute and embarrassed. I look at them as Chantal carries on a conversation. They are as much a part of the château as its towers and walls and courtyard, as much a part of Gageac as the cemetery with its tall cypress trees, the church and the cross at the bottom of our road. I want to know more about them all.

The warmth of Monsieur and Madame de la Verrie's welcome has been humbling. We stand to leave and are led into the courtyard again with its silent beauty where they bid us farewell.

We stroll back along the road to the house. To the left of us is our garden. It is a three-acre wilderness. In fact, through the haze of the midday sun it, too, looks rather beautiful; a riot of campanula and poppies, wild iris and teasels. At the bottom of the field is a row of artichokes that has grown to gigantic proportions, pale silvery green crowns with purple centres.

Our visit to the château fresh in my mind, I want to explore our surroundings a little; I want to know more of the people who have lived here in the past.

Between the church and our house is the monument du mort, the war memorial. Like thousands of others in villages the length and breadth of France, the four corners of its base are stone representations of cannon shells. They are linked by an iron chain that encloses an obelisk, topped by the crossed swords insignia of the *Combattants de France*. At the base of the column are listed those who lost their lives in the first and second world wars, twenty five for the 1914–18 war, one for the 1939–45, one also for the Algerian war. Their names are carved baldly in the stone; no indication of their ages or where they died. But for most of them it will have been in the mud of Flanders, a long way from the sun and warmth of Gageac.

I walk past the church and the monument to the bottom of the road where a large, wrought iron cross stands on a stone plinth. And then left towards the cemetery, the cypress trees dominating the near horizon. They stand behind the walls of the cemetery, brooding and peaceful.

Entering the cemetery through the iron gates, the stillness that I found in the courtyard of the château returns. The cypress trees are immense, towering over the graves; sentinels of peace and protection. The dead are laid to rest by families; Famille Queyrou, Famille Prevot, Famille Tessandier. Some family plots have stone crosses as ornaments, some small wrought iron enclosures, while others have large, marble slabs on which the family names are engraved.

Reminders of France's twentieth century wars are everywhere. On some of the gravestones are the names of those listed on the monument next to the church, with *Mort pour la France* inscribed on their plaques. Others carry salutations from old comrades associations. Some graves have only a wooden cross.

A large stone angel kneels over the Truchasson family vault, while the simple grave of Monsieur Cholet carries a small plaque indicating that he was a prisoner of war. And Gabriel Devaux, the sole casualty of the Algerian war, looks out of a photograph that was probably taken on his eighteenth birthday. An English name amongst them, the Hawkins family, highlights the presence of the English in these lands for centuries past ever since the

Hundred Years War. Further along in the far corner is a simple tomb and on it lies a small, stone coffin. I read with a frisson, despite the heat: 'Suzanne de la Verrie de Vivans, daughter of Pierre and Madelaine, who died, aged one.' Later I learn she drowned in the small pond within the walls of the château.

I close the gates behind me and walk back towards the church. A cypress tree stands next to it. The door is open and I enter, eyes momentarily unaccustomed to the dark after the dazzling sun. It is cool and silent. Chandeliers hang from the ceiling, strangely out of character with the simplicity of the church. The altar has two large, wooden angels kneeling at either side of it. To the right of the altar is a statue of Our Lady and to the left, a large statue of Joan of Arc, brandishing a flag and her sword and dressed for battle.

I sit down on one of the chairs for a moment, a path of sunlight from the open door reaching almost to the altar. Dust is captured in its light, sparkling and dancing. The silence is broken by the sound of a tractor in the distance as I walk back along the carpet of light towards the door and the heat of the sun outside.

Chantal's stay passes all too quickly, as does Sophie's, my step-daughter, who came for a short three-day break before going off to Boston. 'Careful, Soph, Mum'll rope you into visiting the neighbours,' Chantal warns her before leaving for England and preparations for college. 'French and flowers are what you need.'

James has gone again as he has business in England.

'Come to lunch, darling. You're on your own and there are some nice people I'd like you to meet.' A welcome invitation has come from our French friend, Rosemarie Doughty. 'Don't worry darling, Odile speaks English and you can practise your French.'

Odile, the epitome of a French woman of a certain style, is beautiful, with a deep, gravelly voice due to the consumption of at least seventy cigarettes a day, she is strident and forceful with great generosity. At lunch, she speaks at what seems like humanly impossible speed, and my head hurts with the concentration of trying to comprehend. Periodically, I give up the struggle, and slip into reveries. I imagine trying to speak English as quickly as Odile and the others are speaking French and simply cannot. I imagine surprising them all by suddenly being able to speak French fluently and elegantly, understanding everything and using the subjunctive with style and panache. '*Il faut que j'y aille*' – I must go now.

Please let the meal end, please let them not ask me another question. I must learn French.

'I am learning English. I have a teacher,' she is saying. 'He is English and his wife is French. Do you know them?' She continues: 'You must come and have dinner at my house.'

The phone is ringing. Of all the stressful features of my first months in France, the most dreadful is the telephone. The prospect that there might be a French voice on the other end of the line requiring a response is too awful to contemplate. If James is in England, I have no option but to answer it myself. When he is here, he can be counted on to answer it and speak some French to the caller, and he also answers the door when someone knocks. Which is the next most stressful situation I can imagine. Our initial experience with Pepita is one I certainly don't want to repeat, and absolutely not on my own.

Now I am on my own and the telephone is ringing.

'*Madame, je suis votre voisin. Votre chien est en train de détruire ma fenêtre. Il faut venir le chercher!*'

Gilles Cholet has to repeat his message in a number of different ways before I understand that Sam is destroying one of his outhouse windows in a frantic attempt to reach one of his bitches. He wants me to come and retrieve him. A one-sided discussion takes place while he tries to give directions to his house. My heart is racing again. It is dark and I had been concentrating on 'Leçon 2' of *Au Courant*, '*Les Étrangers en France*', feeling that I am at last making some progress. The ensuing non-conversation only serves to underline that this is pure fantasy.

I set off in the direction of the château. I have at least understood the words 'château' and 'droit' amidst the torrent of words. He certainly sounded pretty angry. He shouted the directions. Probably he is even more frustrated at not being understood than with the amorous intentions of Sam.

I call out Sam's name into the dark, hoping against hope that he will appear without me having to spend the night wandering around looking for an unknown house belonging to a nameless neighbour. To my relief, he appears, wagging his tail and apparently pleased to see me. Not knowing whether to hug him or scream at him, I grab his collar and turn back towards home and safety.

'*Heuh! Madame!*'

My heart sinks; what am I going to say? Nothing is the short answer. In my head I had rehearsed '*Bonjour, je m'appelle Madame Atkinson. Je suis désolée pour les chiens. Je suis désolée pour les fenêtres. Je suis désolée . . .*' I turn to see, for the first time, Gilles Cholet.

'*Salut. Gilles Cholet. Il est beau, votre chien. Comment s'appelle t-il?*'

Short, tanned and handsome, he extends his hand and smiles; a broad, wide smile. I take his hand and mutter that I am, as usual, *désolée* and that Sam is called Sam. He talks to me with an apparent confidence that I can understand and, to my surprise, I can decipher a few words among the torrent. His accent is unlike that of Madame de la Verrie's, Rosemarie's or Odile's. This is the local accent that I would come to know so well and even use myself from time to time.

My contribution to the conversation includes '*deux, l'autre est Luke*' interspersed with many a '*désolée*'.

'*Ce n'est pas grave. Mais, a mon avis, il faut les enclôturer. Bonne soirée, Madame. Bonsoir, Sam.*'

Yes it is bloody grave, Sam. And what does *enclôturer* mean? As we walk back to the house, the white owl screeches and flies from its perch in the lime tree.

'We've got to fence in the garden to keep the dogs in,' I say to James when he returns. 'Three acres is not bad for a run.'

He is definitely not in agreement. 'No, it will spoil the look of it. And anyway, we're going to plant vines in some of it.' The look of the garden is not something I feel needs protection on aesthetic grounds. 'The dogs will settle down, don't worry,' he assures me. They don't.

'*Bonsoir, je suis contente de te voir,*' says Odile in her deep voice, smiling and welcoming me into her home. I have eventually found her house, after losing my way twice. It is twenty-five minutes by car from me, at Pinheilh, near Ste Foy la Grande. Her daughter, Peggy, and son, Thomas are there, and I am feeling distinctly unrelaxed.

'*Salut,*' says Thomas.

'*Bonsoir, ravie de te rencontrer,*' says Peggy.

'Mum, can't you speak a little more slowly?' says Peggy, as Odile rattles out words and phrases that hit the outside of my brain and stay there, refusing to transform themselves into anything comprehensible. I am overwhelmed with gratitude towards Peggy.

The evening passes with some English, which all three can speak, but mostly French is spoken. The topics centre on *les Français* and *les Anglais*, food, wine and customs. My by now overloaded brain has given up concentrating on the language, and I find that this is a very good thing. I am beginning to understand some of what is being said.

Relaxing over coffee, and perhaps helped by the wine, I string sentences together and seem to be understood. Or maybe I only think I am because of the wine. By the end of the evening, I've agreed to return the following week. I drive back home, exhilarated by the achievement of passing an evening in a French home and actually being able to participate. As I get out of the car, the white owl continues with its deep breathing exercises.

The dogs are a continuing nightmare. They wait until I'm on my own to wreak havoc on the neighbourhood. A large woman in overall, socks, slippers and sabots, appears at the door. She is the wife of Jean de la Verrie's worker, Victor, and her name is Guillarmine. The dogs have rampaged over her garden, she says, leaving their paw marks everywhere. Her seeds haven't taken yet, but never will if they are trodden down by their large, clumping feet. Pepita arrives at the door with another package, loosely wrapped in an old edition of *Sud Ouest*, the local paper. I can guess . . . Gilles Cholet rings and now I know what *enclôturer* means – a fairly radical way to get to know the neighbours.

I decide that the garden will be enclosed with stakes and wire in spite of James's objections. This need only be temporary, until the vines are planted, I tell myself. I'll tell the same thing to James this evening by telephone. Boujema arrives and I explain in sign language what must happen.

The following day, *enclôturer* is the name of the game. The dogs gallop around the perimeter chasing each other and leaping up and down in the mad way that setters do. They seem positively delighted with the new system and I wonder why I have waited so long to do it.

Gageac has a *cantonnier*, a town council worker. Always present, he is part of the place, looking after everything. He cuts the hedge around the mare's field, he cuts the grass around the church, he lights bonfires, and he cleans ditches. His name is Michel Founaud. I see him daily, working in and around the commune. He usually nods, greets me with '*bonjour*', then drops his head and continues with his work.

'*Bonjour, Madame,*' he says, hardly looking at me as he cuts the hedge opposite the house one morning. In his late forties, he has large, pool-like eyes, a large nose and a liquorice-papered cigarette hanging from his mouth. He is dressed in faded baggy shorts, reaching to his knees and held up with string, and a vest. On his head is a precariously balanced handkerchief, knotted at its four edges.

'*Bonjour, Monsieur,*' I reply.

He avoids my eyes and carries on hacking at the hedge. '*Votre fosse est bloquée,*' he says, jerking his head in the direction of the garden.

I have no idea what a *fosse* is and murmur *oui*, smiling; it's no doubt good to have one *bloquée*. I quickly discover it isn't. He explains, and I follow his meaning as well as I can. The ditch, for that is what it is, is blocked and the entire residue from the kitchen, the bathroom and the washing machine are accumulating in the water pipes of the house. In two days or so it will overflow onto the floor of the workshop, which houses, among other things, the washing machine. The inevitability and timing are somehow evident to Michel as he stands by the hedge, calmly hacking and pulling.

He puts a machine on the back of his tractor, and without a word sets to unblocking the *fosse*. '*Eh voilà,*' he says, when he's finished. He smiles and wipes his hands on his shorts. He talks in short bursts of incomprehensible French and mostly looks down at his feet as he does so, occasionally raising his head to give a smile that lights up his face. 'Okay, bye,' he waves, and is gone.

I watch him as he drives his tractor down the road. He stops at the cemetery, dismounts and disappears inside its gates, then minutes later reappears, leaving behind a thin ribbon of smoke where he has lit a bonfire to burn the dried and dead flowers that he gathered up there earlier in the week. He is silhouetted against the wall and the cypress trees as he waves again and is gone.

Chapter 3

THE SUMMER OF 1990 IS PARTICULARLY HOT AND DRY. WARM, BALMY evenings with suppers in the courtyard; hot, sunny days, spent working on the house, but also relaxing and enjoying the sun.

The sunsets are spectacular. Night after night of deep magenta skies, dramatic against the backdrop of the château. The tinny perfume of geranium, the heady scent of lavender and rosemary, and the honeysuckle, almost overpowering, create an ambience of sun-drenched peace and happiness, southern heat with exotic overtones.

James spends a lot of time in the *chai*, the cellar that will house the wine, preparing it, with the help of Richard, and sometimes Charles, for the forthcoming harvest. Charles and Laura have been over again to stay. Charles has begun to wind down his business in England, and has more spare time. His energy is boundless. While they are here they talk of buying somewhere in France. 'You'll have to put me to work if we move here,' he says. 'Can't be doing nothing.'

By August, the grapes have plumped out into large red bunches. Boujema arrives with the tractor for one last spray before his annual holidays. Richard explains that from now on, one simply waits for ripening. No more spraying, no more cutting; a gentle relaxed period.

I have become used to being surrounded by vines, particularly those around the house or, more specifically, those next to the washing line. I inspect them each time I hang the washing out or take it in. There are five rows between the washing line and the road; small rows, intimate and approachable. I discover they are sauvignon grapes one morning when a man passing on his tractor stops to chat.

'*Bonjour, Madame. Jean de la Verrie,*' he says, dismounting and striding over to shake my hand. His hand is a large worker's hand. In his late fifties, he is of medium height with large, gentle eyes, a deep voice and a tanned, handsome face. I have seen him before, driving past the house on his tractor; indeed Laura and I soon dubbed him the 'Handsome Monsieur de la Verrie'. One of the sons of the Comte and Comtesse at the château, he lives in the next most beautiful house in Gageac after the château. Jean Brun is a fifteenth century manor of tall, sloping rooftops with elegant but simple proportions and gently rolling lawns.

Jean de la Verrie works the vines around the château of Gageac and Jean Brun. He often waves and tips his hat when passing, but now is the first time we have spoken. He removes his hat. '*Ça avance, le muraison*'. The ripening is developing.

I imagine he thinks I know more about vines and grapes than I do. I nod and hope he won't ask me a question.

He doesn't. He talks gently in a deep voice with a hint of laughter in it. He talks of the year we are having, the heat, the *mûraison* of the grapes, the forthcoming vendange. I listen, captivated by his voice and surprised I can understand some of what he is saying. '*Elles sont précoces cette année.*'

Precocious grapes! My mind grapples with the concept. '*Oui, oui,*' I stutter.

'It's often the case with sauvignon' he says. 'But each year is different. One never knows – time will tell.' He holds out his hand again, shakes mine, and is gone.

Serge Chassaigne, the builder, continues his work on the house and the roof. His family is one of the oldest in the region. His brother, Philip, is a mason like him, as was his father before he retired. His unlimited patience with my attempts to make myself understood, whether in a crisis or making a simple request, is extraordinary. With his gentle voice and calm manner, he arrives each morning with his two workers. We have a mutual understanding, even though I can't communicate with him in French. Like Michel, he is integral to the life of the village and my life here.

'Come quick, there's a duvet in the workshop!' I shout as they arrive for work one morning.

He looks at me in his kind, steady way, turns to his workers and repeats in his slow drawl: 'A duvet. *Ah bon.* Come on men, there's a duvet in the workshop. *On y va.*'

I had gone into the atelier for a brush to rescue another stag beetle from the kitchen floor to be astonished by a white owl, possibly the same owl that lives in the lime tree, which had somehow got trapped in there. It flew at me in fright. The feeling was mutual.

'Ah, an owl,' says Serge and lifts it down from its perching place.

Couette is duvet, *chouette* is owl, I mutter to myself as I depart in the usual state of embarrassment at my dismal French.

The vendange is approaching. Summer days are still with us, but one senses the end of the season. The mornings are fresher with a subtle change in the quality of light. There is a serene beauty in the gentle contours of the rows of vines, their surge of growth over for the season. They are still a mass of green leaves, but with shades of yellow and pale reds, a spectrum of delicately changing colours.

Boujema arrives one afternoon. There are two rows of eating

grapes among the vines at the top, next to Roger and Pepita's. Can he take them? he asks. No good mixing them with the wine grapes. And would we like some picked for us? A bucket-full is deposited on the table in the courtyard. Large, black and perfumed, they are full of flavour and juice. I'm surprised at how different they are in size, shape and general aspect from the ones I see around the washing line. Does this mean the others aren't ready? How fantastic it is that we can produce such things with so little effort. A bit of spraying here, bit of mowing there . . .

'You're probably vendanging next week,' announces Richard, who is going to help us with our first harvest. 'We might as well use the same *vendangeurs* for both properties.' Richard has fourteen hectares of vines compared to our little patch of four and a half. That is to say, his is a serious business. As Richard is starting a week before us, tomorrow in fact, I am requisitioned to pick at his vineyard, along with Chantal, who is over for the last week or so before the new term, Andrew, my stepson and various friends of theirs who are also staying.

'But we don't know what to do. What if we make a mistake and cut off the wrong things?' we ask.

Richard laughs and assures us that the only wrong things to cut off would be our fingers and that self-preservation will probably ensure their safety. 'Come over before eight,' he says, 'and I'll show you. We start on the dot at 8 am.'

Gageac has been transformed overnight. Gone is the peaceful, sleepy village. Hardly even a village, given that the commune, of which our house is the epicentre, consists of a château, a church and not more than six houses. Gone is the feeling of an eternal, sun-drenched holiday. Gone the sense that no one really lives here except for us.

Tractors thunder by pulling empty trailers, which reverberate as they bounce over the bumpy road. Cars follow their cloudy wake of dust, drivers intent on the business in hand. Monsieur de la Verrie passes at great speed on his tractor, tipping his hat. His worker, Victor, on another tractor, which pulls a huge, cream-coloured machine, follows. Behind him is his wife, Guillarmine, on her bicycle; the lady whose garden and seedlings Sam and Luke destroyed. Yet another tractor, this time pulling an open wooden trailer, on which is perched what looks like half an enormous wooden barrel, rushes by. It is driven by a man I have never seen before, an elderly, solid man wearing pale blue trousers and jacket, with twinkling eyes and a dark blue beret. It is only 7.30 am for God's sake! He raises a hand in acknowledgement as he passes, eyes firmly fixed on the road and the middle distance.

'Where have all these people come from?' asks Chantal.

A sense of urgency, work, concentration and determination has entered Gageac. Gulping down coffee and gathering what we think are necessities, bikini top for one of Chantal's friends, cassette player for Andrew, and a croissant or two for me, we pile into the Citroën Diane and drive the three kilometres to Richard's.

Twelve or so people are milling around. On the ground are twenty brown plastic buckets, a wooden box in which sit a pile of grape cutting scissors and, nearby, the tractor with obligatory trailer. Boujema sits on the tractor. '*Bonjour!*'

He laughs, happy to be starting the vendange. '*On y va?*' he says to Richard. We feel intimidated and faintly embarrassed to be there, a group of foreigners dressed for an outing to the sea, unlike the *vendangeurs* in front of us, *bleus de travail*, hats and not a cassette player, bikini top or croissant in sight. They each shake hands with us and wander off toward the vines. We follow, looking around desperately for Richard and some insight into what is expected of us. He appears again, smiling, laughing, obviously equally

37

delighted to be starting the vendange and also happy to have six or so additional workers in us.

We are picking white grapes. 'Work in pairs, one each side of the vine. Don't leave the vine until all grapes have been picked, don't change rows, and if you finish your row before your partner, join his row at the end and work towards him.' But how do we cut them? 'Look into the vine and search for bunches, then cut the stalk and make sure all the grapes fall into your bucket. Place the bucket under the vine in case you drop a bunch. We don't want to lose a single grape! Let's go everyone.'

We gaze into the vine, through the mass of leaves, and find the bunches. Forgetting to place the bucket where I'm told to, my first bunch falls onto the ground, grapes bouncing off. I scrabble to retrieve them, hoping no-one has noticed.

'Don't put your scissors in the bucket!' shouts Boujema. Why, I wonder, hastily taking them out of the bucket, where I've dropped them while I gather the now completely disintegrated bunch.

'Never put them in the bucket,' explains Richard, 'not for an instant, because you'll either lose them among the grapes, or Boujema may take your full bucket to tip into the trailer while you're doing something else and it will break the press!'

Horrified that I might have committed such a wanton act, but unable to imagine what other thing I might be doing while Boujema creeps up and steals my bucket, I firmly grasp the scissors. It's all I can do to cut a bunch and place it in the basket, for God's sake! Place the bucket under the vine and start again. I peer into the vine. The grapes have gone.

My partner (Richard has sensibly separated the novice workers from each other and placed each of us opposite a real worker) has long ago denuded my vine of its fruit. While I've been scrambling to gather one collapsing bunch, he has worked the next twenty vines, both on his side and on mine. I rush to join him and recommence in earnest.

The atmosphere is jovial and the real workers chatter away, laughing and shouting out to each other. They relate stories, or tell jokes, or sing. Every so often someone shouts '*panier*' and Boujema appears, cigarette in mouth, and empties the basket into the large trailer. I quickly understand that my *panier* can indeed be stolen whilst I am in mid pick, so to speak. In an instant it's gone, then back, empty and ready to be filled up again, exactly where I had left it. Not only does Boujema do this effortlessly for all the twenty buckets, but he wanders along behind me and others, picking odd bunches we have left behind.

A rhythm is established. 'This is your first time?' asks my partner.

'*Oui*,' I reply, ashamed not to be able to chatter like the rest of the team.

Periods of chatter and laughter are occasionally replaced by periods of silence, a sort of communal concentration, then again by chatter, laughter, a song that most people join in with, a joke that everyone laughs at.

My concentration wavers. I have almost cut my finger off, or so it seems. I squeal in fright, seeing first white skin (did I really see bone?), then blood, a deep bloody cut, accompanied by stinging pain.

'Ah, don't worry, Madame, the sweetness of the grape will save your finger and heal the wound,' says my partner, unconcerned, but kindly.

'*Ah la, la*,' laughs one of the women workers, 'now you're a real worker!'

'Never mind the sweetness of the grape, René!' suggests one of the other women amidst raucous laughter, 'kiss it better – that'll help!'

It is hot and I wish I had brought a hat. The croissants have long ago dried up. There isn't a second to spare to eat them as I must keep up with my partner and anyway, no one is stopping to eat. I wish, too, that I had a tee-shirt that covered more of my back, or, more specifically, my neck. At least my finger has stopped bleeding. My hands are sticky with the sugariness of the grapes, the scissors are seizing up because of it and the backs of my legs are burning.

I look back towards my team, hoping that they have more protection from the sun than me. I see and feel their misery. Chantal appears to be picking entirely on her own, although closer scrutiny reveals she's still on her original row. Her partner, having finished working up his side, is calmly working his way back down towards her. She is looking very hot, very tired and very bothered. Her friends are in a similar state. Andrew is soldiering on, at least with a hat on his head, but he has taken off his tee-shirt completely. I see that his back has turned a deeper shade of pink – a very deep shade in fact. In the next row, another of Chantal's friends, a rather beautiful, pale-skinned Irish colleen is no longer so pale.

Midday mercifully arrives and we re-group. Andrew is told to put on his tee-shirt immediately. He objects, saying his back doesn't hurt and anyway, he's only here for a few days and wants a tan. The Irish colleen is beginning to feel the effects of the sun and agrees that she has changed colour. It becomes evident that two of our party are not here.

'Becky's back couldn't take it. She's lying down on the floor of Richard's atelier,' says Chantal, 'and has been since 9 o'clock. I've got to take her back home. She's in tears with the pain. And Maria's gone to find a plaster for her finger and a hat. Mum, let's face it, we're useless.'

With reluctance, I have to agree.

'We'll have lunch at home, Mum,' says Chantal. 'And cold showers!'

Andrew, at least, is prepared to stay.

'We'll come back and get you at six,' shouts Chantal as they pile into the Diane with relief.

For Andrew and me, initial enthusiasm has been replaced by dogged determination. We battle on, hour after hour. If it was hot this morning, it is roasting this afternoon. I change sides to relieve the hot, prickly sensation on the backs of my legs and neck from the strong sun. Someone kindly lends me a hat. After a modicum of protest, I accept gratefully.

The stickiness of the grapes increases with the heat of the sun. '*Oh là là, ça colle,*' quips one of the workers. '*Magnifique, combien ça pese?*' I reflect on this. Why should we need to know how much a grape weighs? And why is it magnificent that they are sticky?

My mind is dulled by the heat, the food digesting in my stomach, the dull drone of the tractor, and the desultory conversations amongst the rows. I pick, half-listening to the sounds of French being spoken. Everyone seems calmer this afternoon. The voices are gentler, rhythmic, slower even. I half listen, and realise that there's a melody running through their voices and the language. I am beginning to recognise it and decipher whole phrases. Some of the words that have formed part of my vocabulary homework are popping up, but in context, in sentences.

I am suddenly awake and alert, listening, translating what I hear and euphoric with the unexpected progress I've made. I can't understand everything, but what seem like great chunks of language are suddenly comprehensible. I want to tell someone, shout that I am, in fact, understanding some of what they are saying, even if I can't yet reply; understanding in a real, connected way rather than odd words here and there. Andrew is too far away for me to shout it out to him, and Richard is nowhere is sight. I must tell someone right now. No, there's no time. Just keep listening, I tell myself.

And I am. And it's working! I am concentrating, and listening, and hearing the melody and deciphering the words. The speed at which I'm working quickens. I must keep up with them all. Must hear what they're saying, particularly that woman there, whom

I can understand much more easily than the others. '*Ooh la, regarde! Elle va nous dépasser, tellement elle travaille!*' she exclaims jokingly. She's going to overtake us she's working so hard. I want to shout *I can understand! You don't know, but I can, honestly!!*

James arrives to pick us up. He had decided not to come and pick. Andrew is by now lobster coloured. My euphoria at the language breakthrough knows no bounds. In spite of swollen legs, a sore neck and an aching body, I'm ecstatic. As I bubble over with enthusiasm, relating the story of my accomplishment, I scratch behind my knee, an itch that's no doubt due to being sunburnt. Andrew is quiet in the back of the car, suffering from the effects of the sun and a day's hard work.

'The girls are in a terrible state,' says James. 'Becky's lying on a bed with backache. Chantal, Maria and Sinaed are sunburnt, one of them can hardly walk and they're all complaining of swollen lumps and itching on their bodies.'

'I'm pretty itchy in certain places actually,' says Andrew from the back seat.

And indeed, so am I. In fact I'm very itchy and as I look at the spot on the back of my knee it erupts into a large, white bubble. The other leg is in the same state, along with my waistline, crutch, armpits, inner thigh and just about every other warm spot that has a fold in the skin. The sunburn, the tiredness, the hunger, worry for the girls and Andrew, and the fact that it's my fault for making them vendange in the first place, deflate my euphoria.

James is not exaggerating. Chantal and the girls are in a parlous state. They are scratching, as we all are; crutch, waist, backs of legs, under breasts. The pale Irish colleen is now a unified pink one, Becky is prostrate on the bed, in obvious pain. Maria's finger is pretty swollen and she can hardly walk, having stood on a nail in the vines.

Chantal's skin has reacted more violently to the white bubble syndrome than the rest of us and some of the lesions are weeping and angry-looking. I ring up Richard.

'Ah yes,' he says. 'Forgot to tell you about them. *Aouta*. Have you got them? They're called harvest bugs in England. Little mites that climb up to any warm place and lay eggs in you. It's the eggs hatching that's causing the discomfort.'

We are collectively horrified.

'Can't avoid them at this time of year,' he says. ' If they're out, they'll get you. Can't treat preventively either. But once you've got them, you can get something from the chemist to paint on them. It does help a bit. But if I were you, I'd just put alcohol or eau de cologne on them. Buggers, aren't they? I'm scratching like mad too.'

If the girls were in a state before knowing this, they are now stricken with disgust and despair. 'Mum, why didn't you tell us we were going to be invaded by these bloody things? Disgusting!' Chantal screams. 'What are we going to do? How can we get rid of them? We're not going back tomorrow, Mum!'

Some benzochlorine from the chemist and more showers bring some relief. After dinner, sitting around the table under the lime tree, we feel shell-shocked, but calmer and better than we've felt all day. Surrounded by vines, whose serene sculptured forms belie the threat they pose as harbourers of the ferocious *aouta*, we watch the spectacular sunset over the roof of the château.

'Mum, you don't mind if we don't come back tomorrow, do you? We just can't.' says Chantal quietly.

'I'll come,' says Andrew stoically.

Gageac has become a veritable industrial estate. Tractors rush by, trailers in tow. Monsieur de la Verrie's *chai*, which is at the bottom of the road, just past the derelict house, is obviously in full use. There are pipes running from a trailer to a hole in the ground. His worker is standing by the tractor, revving it up. I am not sure why he is standing there, or what Monsieur de la Verrie is doing, shouting directions one second, peering into the hole the next. The man with the lovely smile and large, twinkling eyes who I saw for the first time yesterday zooms past on his tractor again this morning, waving again, intent on the business in hand.

No time for speculation as Andrew and I are off for another day of vendanging. My white bubbles have erupted in the night, giving me hell, and some of them are still weeping. Andrew's back looks horrendously raw and the backs of my legs are only slightly less swollen than they were last night. The morning sun is already giving out heat as we drive over to Richard's.

We are greeted effusively by the workers. They weren't expecting us, they exclaim. Welcome back! How's the back, Andrew? How's your finger, Patricia? Has René's charm worked? Shall he be your partner today? No scissors in the *panier*, huh? We've brought you a hat. Where's your *bleu de travail*, Andrew?

Richard is delighted. We have not let the side down. We are affiliated workers, affiliated Frenchmen.

A week later, and I consider myself a seasoned picker. I've worked through the pain barrier and the *aouta* and my swollen legs are under control. Regretfully I say goodbye to Andrew, Chantal and the other girls. They, too, are recovered. Andrew's back is no longer deep magenta. Sinaed's skin, if not what it was before, is no longer shocking. Becky has recovered from her back problems and Maria can walk again without limping. Chantal's weeping sores have dried up and, all in all, they claim to have had a good time.

'I'm sorry I was so useless,' says Becky. 'Edge will be much better at it next year, really.' Edge is Becky's brother, who wants to come and vendange next year.

'Yeah, really sorry to be such weaklings,' adds the Irish colleen. 'Bye!'

Today, the team is moving over to our own vineyard. We will be picking white grapes, in a *parcelle* just five minutes' walk from Richard's vineyard. To my surprise, we finish almost before we have started. One hour and the vines are devoid of grapes.

James hasn't come over to help with the first pick of our own grapes. In fact, he hasn't picked at all. He has things to do in the *chai* and the sun is too strong for him. He has never liked direct sun much, which is one reason we used to come to France in spring rather than summer, so it's odd, really, that he has chosen vines and this part of France as his dream.

Back to Richard's vines and we start to pick red grapes. As I pick, I wonder whether James is okay. He was ill a year ago in England, just after he left the bank, with what was diagnosed as probably a rheumatic virus. His symptoms were fatigue, depression and painful joints. The doctor said it would go away, which it did, and never come back. He certainly seems fine, apart from being bitten by mosquitoes. We laugh as he counts the number of bites he amasses each evening as we sit in the courtyard after the day's work.

I am no longer being bitten by the *aouta*, or perhaps I've become immune to them. Right now, I'm concentrating on picking. Picking the red grapes is a speedier affair — the bunches are more compact, easier to see and quicker to pick. I feel I am now a veritable expert, and I even converse, after a fashion, with the rest of the team, who I now know by name. I have become adept at snipping off the grapes at just the right spot on the stalk. One swift cut and the bunch falls neatly into my bucket. I no longer need to place it there carefully or look as I do so; it hardly touches my hand and the next bunch is cut before the preceding one hits the bucket.

As I arrive back home, James is grinning widely, standing by the press. Its barrel is turning, slowly pressing the white grapes picked earlier in the day and transported over by Boujema with the tractor. Juice is gushing into the huge tray underneath the barrel of the press. 'Taste it!' says James, ecstatic. He scoops some juice out of the tray and into a glass and I taste, for the first time, white grape juice, fresh from the press. Our very own grape juice. It tastes tart, but fresh and cool.

Each time the tray fills up with juice, the overflow is sucked through an adjoining pipe by a pump next to the press. Another pipe sends it to a huge vat, some twenty feet high. The noise is deafening, what with the lumbering press and the pump and I now understand at least why Monsieur de la Verrie was shouting and gesticulating to his worker outside his *chai* last week as they probably couldn't hear each other.

The *chai* is a mess. After pressing, the stalks and skins are tipped out onto the floor. With the aid of a huge brush and an equally large shovel, we throw the skins and stalks into wheelbarrows and take them to the roadside. By law, we are not allowed to keep them, Richard has told us. They represent a form of government tax. Why would we want to keep the grape skins? I ask myself as I shovel them into the wheelbarrow. The distillery will send along a lorry to pick them up and some of it will be used to make brandy. I look at the small mountain of skins at the side of the road and wonder how these will be transformed into wine.

Astonished at the quantity of grape skins produced from our morning's work, I try to imagine what it must be like over at Richard's, where there is a vastly greater quantity of grapes. Does he really shovel all that stuff into wheelbarrows as we have just done? I'm tired, hot and hungry. While I fill the barrow with skins and wheel it through the courtyard, and past the washing line to the roadside, James adds yeast to the wine vat, then sets the hose pipe onto the press to start cleaning it.

My hands begin to sting with virgin blisters from the shovel and brush. Hours later, the last load of skins is deposited by the road. I must eat and sleep. The white owl hisses and breathes deeply as I pass the lime tree. Tomorrow I won't be going over to Richard's. He has enough grapes for the moment and will be pressing all day, rather than picking. Thank God.

'Mum!' It's John, my son, ringing from Thailand where he works in information technology. 'How's it going? Is the vendange done?' He wishes he could be in Gageac, he says, to help. He worked a vendange the year before, so knows what it is like. I tell him about Andrew, Chantal and the girls. He laughs and says he'll take a raincheck on the wish being granted. Emma rings too. She is living in France, near Aubusson, but can't come over as she's working.

The industrial estate of Gageac has an ordered rhythm now. The tractors rush by with empty trailers and return an hour or so later full to the brim. The strange cream-coloured contraption attached to the back of one of Jean de la Verrie's tractors is a vendanging machine. It straddles a row of vines, lumbering slowly up and down, driven by his worker. It gathers kilos of grapes deep into its belly. Elizabeth de la Verrie, Jean's wife, follows with a bucket, picking the bunches that remain. Every so often the machine stops at the end of a row, Jean de la Verrie arrives with his tractor and trailer, and piles the load into his trailer. How much easier it would be if we could do that! They gather in kilos in record time, whereas our hand picks are laborious and slow. But Richard has explained that hand picking is more precise. One can cut off any rotten grapes which would otherwise be included in the pick, as well as leaves and other debris.

Further up the road, past Roger and Pepita's, the man with the twinkling eyes and pale blue suit is hand picking with a team of six, nothing like our mammoth team of twenty. The trailer housing the enormous half-wooden barrel is stationary. The pickers move slowly up and down the vines, every so often tipping their *paniers* into the barrel. Next to the cemetery, yet another vendanging machine is slowly progressing up and down the vines, gathering in kilo after kilo of grapes. And the sun sends down its hot, copper rays relentlessly on the workers, the machines and the grapes.

I feel thankful that I am not picking today. However, when James cleaned the press last night he gave it just a cursory rinse, and this morning it needs to be scrubbed thoroughly.

The press consists of a cylindrical wooden barrel sitting in a frame. The interior houses five large hoops with connecting chains. As the barrel revolves, a large centred screw draws the plates at each end towards the centre, then pushes them back outwards to full opening capacity.

The hoops and chains, which work like an accordion, keep the load moving and break up clumps of skins that might block the system. And now it is empty, apart from a million translucent skins hiding in the crevices, on the screw, under the frame and around the chains.

The yeast that James added to the white grape juice last night has started fermenting. I press my ear to the vat and hear the sound of millions of yeast bodies chomping away at the sugar and turning it to alcohol. It bubbles and hisses and James is delighted.

If the press with its cylindrical drum was noisy with a full load yesterday, it is even louder now it is empty as the hoops and chains clash with each other and the machine creaks and lumbers through its repertoire. While the machine is in motion, I spray it with water from the hose to clean off skins and pips. It is astonishing that last night it looked clean, yet this morning it is disgorging skins and pips as if it hadn't been touched.

Soon drenched but by now intent on finishing the job properly, I stop the machine and climb in, crouching down and squeezing myself in beside the enormous screw and between the hoops and chains. I scoop pips and skins into a bucket. The translucent skins are everywhere and the pips hide in clusters. Climbing out again I set to with the hose and water. Still they keep coming. At least the floor will be clean. James had been muttering about how difficult the floor was to clean when he was preparing the *chai* for the vendange.

It soon becomes clear that there is no floor. It is, in fact, mud and stones, or rather mud and pips and skins and stones. Furthermore, in my attempts to clear some of the water and mud and skins, it's evident that I am pushing them all uphill, along with the stones. The drain where the resultant sludge is heading for is under the vat that houses the wine and is almost impossible to get to.

In fact it isn't actually a drain; it is simply a hole into which the water will run and, on a good day, soak down into the mud and soil fifty centimetres below. I'm exhausted and it's only 3 o'clock in the afternoon. The sun shines on the huge pile of grape skins outside and a pungent smell, sour and vinegary, emanates from it. As I approach it, millions of vinegar flies rise up. Suddenly they are everywhere; on the washing and in the house. They alight on the fruit bowl in the kitchen, and swirl in and around the empty wineglasses from lunch and on the plates; they are disgusting.

With some relief I escape the *chai* and reappear at Richard's the following morning to pick more red grapes. Today there are three new workers; Moroccan women. Dressed in multi-coloured trousers and tops, they look exotic. Their hats are beautiful, large straw creations with coloured bobbles hanging from the tassels that surround the brim. Under their hats, their hair is wrapped in scarves and over their trousers and tops they wear jalabis. They shake hands with me. One smiles, lightning flashing across her face. Her mouth is full of steel teeth.

The morning passes quickly, the grapes are snipped, the buckets fill up rapidly and Richard's press is full again. The sun beats down with unrelenting power. This afternoon we will pick our own red grapes at Gageac, says Richard.

At Gageac, the vinegar mountain sits by the road still, and flies swarm in their millions. I am exhausted. I managed to clear the water and skins and pips and stones from the *chai* yesterday, but not until late in the evening. A neat pile of stones outside the *chai* door sits as testimony to the task. This afternoon, my blistered hands and screaming muscles remind me of it every time I take a step or move my arms.

'God! Haven't you rung the distillery yet?' exclaims Richard, looking at the rotting pile. As we speak, a lorry appears, a huge arm on its trailer. It stops at the heaving pile of skins and vinegar flies and two men dismount. '*Bonjour! C'est bien le moment, n'est-ce pas?*' they shout and the arm in the trailer lifts, hovers over the pile, then descends and scoops it up. One of the men gathers the residue with a shovel and, with a wave of his hand, they are gone. The smell and the flies go with them. And my gratitude and relief.

We have two types of red grape at Gageac: the earlier ripening merlot and cabernet sauvignon. The merlot is picked and the grapes arrive at the *chai* door. They are not piled into the press like the white grapes, but sent through an *égrappoir*, or crusher, which liberates the juice by removing the stalks, breaking the skins, and sending them into a vat.

The noise from it is deafening. The grapes pour into the machine, and up, through pump and pipe, into a vat. The grape stalks spew out of the other end and I'm back to wheelbarrow, shovel and brush. At least stalks are lighter than skins and pips. I climb the ladder at the side of the vat and peer in. Looking down in to the base of the huge vat, I see one hell of a lot of red skins, pips, pulp and juice.

For the moment, the red grape picking is over. It will be another two weeks or so before we will pick the cabernet sauvignon grapes. There has been no time for French homework; however, my language skills are improving, forcibly, by spending my days with the vendange team.

The juice from the merlot grapes is now bubbling away. Delighted that we have something in our vats at last, James spends a lot of time in the *chai*. The white is still fermenting. Once a week an *oenologue*, a wine scientist, arrives to check on the vats. His name is Jean Marc Dournel. James chats to him and I'm deeply impressed that he's capable of using the technical vocabulary required. Each week for the last three weeks the *oenologue* has taken away a sample from each of the vats for analysis.

A temporary calm has descended on Gageac. The tractors no longer race by and although the door of Jean de la Verrie's *chai* is permanently open, and work is going on inside, there is no longer the sense of urgency and noise of the last three weeks.

As I open the shutters of the bedroom window, the horse gazes up at me, as she does each morning. Bathed in the tender light of a new day, her field this morning is struck with gold. It is already mid-September, and still the sun shines down. She waits for breakfast and stamps her foot in impatience. I descend and cross the road to feed her with last night's bread and an apple. I contemplate the view.

The landscape has changed. As I look out beyond the long stone building opposite the horse's field, most of the vines are radically denuded of grapes and leaves. One section remains — the cabernet sauvignons. There, the leaves are dense around grapes awaiting harvest. The bunches are sparser than the merlot were, and hang much lower on the vines, in long, pendulous groups.

'*Bonjour Madame, je suis ravi de faire votre connaissance.*' A white-haired man with a pink shirt who I have sometimes seen with Jean de la Verrie approaches me as I stroke the horse's face. '*Geoffroy de la Verrie. Bienvenue a Gageac.*' He shakes my hand, smiling. He is small, perfectly formed with intelligent blue eyes that absolutely match his pale blue jeans. He is tanned and clean – and obviously not a manual worker. He is the brother of Jean de la Verrie. He greets James too, who appears at the door and crosses the road to the horse and us.

He lives in Antibes, he tells us, and has come to Gageac for a day or so to escape the traffic and tourists, to visit his parents and see how his brother's vendange is going. He laughs as he speaks, like his brother – infectious, a lilting voice, full of vitality and zest. He speaks beautifully, clear and articulate and the difference between his French and that of my team in the vines is very evident.

'I see you know Eida,' he says, smiling and looking at the horse. Eida belongs to the château and was once ridden by Jean de la Verrie's children. They are now grown up. 'Eida, too, is a certain age,' he tells us. 'She has obviously imposed her will on you , as she does to everyone,' he continues. 'Does she ask you for supper too? *Ah, Eida, tu es redoutable, comme beaucoup de femmes,*' he says, stroking her face. She listens, then stamps her foot. Geoffroy's face breaks into a grin, then he laughs, an infectious, tinkling laugh. He raises his hand as he turns from us to leave. '*A bientôt j'espère,*' and with a jaunty walk, he heads off towards the château.

The holiday is over. 'First pick of the noble rot tomorrow,' announces Richard. 'Make sure you're over here a bit earlier so I can explain. It's not like the other picks.'

This is an understatement. My newly acquired knowledge of picking is now redundant. I have been diligent in spotting and removing any hint of rot on the white and red bunches, as I had been told that any rot on them will spoil the wine.

'This time, you are looking for rot. See this bunch? See this rot? Pick it. If the bunch doesn't have some rot on it, don't take it.' So I must now do the opposite of what I have been doing. It looks disgusting and I can't imagine what sort of wine it might make. 'Sweet wine,' says Richard.

The grapes stick to us, the bunches fall into the *paniers* and the sun beats down. We are tired and hot, but we pick in silence, concentrating and gathering rotten bunches instead of golden bunches and green bunches. Sometimes we pick golden bunches, but only if they are on the edge of turning rotten. If we miss a rotten bunch, eagle-eyed Boujema is there to gather it in. If we include a green grape in an almost rotten bunch Boujema shouts out in consternation. By the end of the afternoon, we are experts.

The following morning, Richard is standing in front of us, grinning. 'Today you can take all of the grapes, the results were so good yesterday. All rules are off!' He has the level of sugar he needs without sticking exclusively to noble rot bunches. You mean pick all the grapes – with or without rot, we ask in surprise? Having learned our new skill, it's difficult to dispense with it. We look at Boujema for confirmation, who laughs raucously, kicking the foot of a vine as he does so. We set to work. 'You've left some behind!' shouts Boujema at me, as I move on from one vine to the next. 'There's a bottle of wine there!' He deftly collects a half-bunch I have missed. We finish at Richard's and move over to Gageac to pick our own noble rot. The grapes arrive in our *chai* and are pressed immediately.

If the red and white harvest make the press difficult to clean, the noble rot is in a category of its own. And again it is my job to clean it. The skins stick fast to the inside of the press, the pips hide in any crevice and most particularly in the corners of the slats of the barrel, and removing them seems an impossible task. The muddy mixture on the floor of the *chai* refuses to disappear. Each time I sweep it uphill, it sludges back down. Each time I think the press is clean, closer inspection tells me otherwise.

'Make sure you wash the press out properly tonight, because if you don't it'll be ten times worse tomorrow,' warns Richard before he leaves our *chai* to go back to his own.

Two days later and it is time to pick the cabernet sauvignon. — The grapes are smaller, but the bunches are longer; elongated, elegant and, most immediately interesting, easier to pick. The bunches drop cleanly and swiftly into the *paniers* as the *vendangeurs*, knowing the end is in sight, pick with gusto. There is sense of gaiety and it's infectious. It is 2.30 pm when the last bunch is picked. All the *vendangeurs* pick on the final row.

As the last bunch hits the *panier*, a unified roar of laughter rises up. '*Voilà, c'est fini!*' *Paniers* are emptied into the trailer, scissors into the box on the seat of the tractor. We walk back slowly along the rows of vines to the *chai* door. A woman picks a bunch of wild flowers and places it on the grapes in the trailer. '*Le gerbeboade,*' she says. 'It brings good luck.' What is the *gerbeboade*? I ask. Nobody can tell me; some sort of Dionysian rite, I tell myself.

We stand around in groups; some of us wash our hands, some our *paniers*. People are chatting, taking off boots and hats, and gathering possessions. Already the harvest is being tipped into the crusher and sent up into vats by Richard and Boujema. One cycle is over, the other is just beginning. Life moves on and they have already left the picking behind for the next phase. We say goodbye to each other warmly. We have spent the whole of the harvest together and become comrades.

Arriving home with a sense of liberation and achievement, I shower off the last of the vendange, happy to leave it behind. No more crusher, no more washing grape-skins out and pushing mud uphill. No more endless wheelbarrow trips to the road and back. No more *aouta*.

Gageac is in the last throes of the vendange. Jean de la Verrie is bringing in his final loads, his worker, Victor, standing next to the tractor and trailer. Jean shouts and gesticulates until the loads are emptied, then drives back up the rows again with the vendanging machine for the few last picks. By the end of the day, the tractors and trailers are quiet. The road in front of the house is strewn with dry mud and grass from the tractor wheels, and odd bunches of grapes that have fallen from the trailers, but it becomes the small country road it was before.

Gageac looks beautiful. The landscape has changed and the rows of green vines are now paths of red and gold. The large leaves of the mulberry trees that line the road leading down to the château have a golden hue against the deep, dark green of the cypress trees in the cemetery. The season is casting its spell, creating a landscape of indescribable beauty. Brilliant colours caress the vines, the trees and the ridge, lending a mystical splendour. The days are still hot and sunny, but the light is sharper and the evenings are fresher.

We are startled awake by the sound of church bells outside the bedroom window. What time is it, for God's sake? It is 7 am and the bell rings insistently. It chimes three times, then is silent, then three times again. After a period, it chimes unremittingly, then reverts to peals of three, after which there is a silence. I leap out of bed and open the shutters. The elderly man with the twinkling eyes emerges from the door of the church, looking solemn. His bicycle is leaning against the church wall. He closes the door and locks it with the large key, beret in hand. He looks at his beret for a moment, shakes his head, then puts it on slowly.

He looks up and waves at me, then gets on his bicycle and rides off in the direction of Roger and Pepita's. At midday, he is back to ring the bells again. *'C'est bien triste,'* he says. He is Monsieur Cazin and he is the official church bellringer. Someone in the village has died, hence the ringing – peals of three for a man, two for a woman. I wonder who decided this rule? Injustice to dead women, I tell myself.

Monsieur Cazin's wife appears beside him. I have seen her often, cycling up and down our road, often with small milk churns balanced on the handlebars of her bicycle. She cycles slowly and laboriously, eyes fixed in front of her, intent on her destination, which is somewhere near the château.

She talks with a slow, clear voice, peering intently and listening with apparent interest to any response. She explains that the man who has died was one of the oldest in the village, apart from Monsieur le Comte at the château, of course. And then there's Madame Queyrou's mother who's still alive, and she's older than the lot of them. But then, she's a woman, which is different. Monsieur Coq's also getting on a bit and as for Monsieur Briand, he goes on forever.

I wonder where all these people live, as there are only six houses or so in the vicinity, but feel flattered that I am the recipient of this information, or any information come to that. I nod and agree. She must go and milk the cows, she continues. Do I like fresh milk? Not everyone does and it's non-pasteurised so she can't sell it but she will leave some outside my door for me if I would like some for the cats we've now acquired or the dogs. She knows we have both and that they like it because they all come to her house for it. Yes, Sam was there just this morning. Oh God, another visit from Pepita is a dead cert.

The milk that Madame Cazin deposits each evening outside the kitchen door has a cap of cream three inches thick . It arrives in a two-litre churn, still warm. She dismounts from her bike, which is placed carefully beside the wall, and slowly and purposefully unhooks our churn from the others. We are obviously first on the run. She talks to the dogs through the fence in a gentle, calm manner as the cats, Edward and Lulu, wrap themselves around her feet. She talks to them, bends down to stroke them, then remounts her bike and cycles off. She too wears sabots. They are obviously *de rigueur* for women around here. Even Jean de la Verrie's wife, Elizabeth, wears them.

A week later, and I have sabots too. The weather breaks and it rains. I can't remember it raining since we arrived five months ago, but this makes up for it dramatically. It rains for what seems like weeks. The dogs wander in and out of the house, leaving mud in their wake. We light a fire for the first time. The heat disappears up the chimney and although it looks charming, it is singularly ineffective. We spend most of the time going to and from the woodpile, which is behind the house. Here sabots come in particularly useful.

Smoke billows out into the kitchen. Serge, the builder, suggests raising the grate and elongating the chimney, both of which we try, neither of which work. Even worse, each time the fire is lit, smoke somehow also reaches our bedroom, winding its way through beams and ceilings, making us reluctant to sleep there for fear of asphyxiation in the night. Serge's final suggestion is to give up on the romantic ideal of an open fire and install a wood burning stove, which we agree to instantly.

The wine in the *chai* does what it does with the care and control of James and the advice of our new *oenologue*, Bruno Bilancini. Jean Marc Dournel's sector has been split in two and Bruno, a new employee at the laboratory, is now covering our sector. Small, with a quiet but decisive voice and a kind, intelligent face, he has come from Bordeaux, and was previously working for the Cave Co-operative at Monbazillac.

Richard also comes over occasionally to witness nature taking its course. The fermentations are apparently finished and there are things to do. James has bought a number of books on wine-making methods, both in French and English, and spends almost as much time on them as I do on *Au Courant* and Marcel Pagnol's *Souvenirs de mes Enfances*.

Serge's work on the house also continues, along with plans to improve the *chai* before the next vendange. A concrete floor would be a significant step forward, not to mention an effective drain. And perhaps an extension to the existing building, which has a shed that is so dilapidated it is barely attached. Serge will draw up plans and we can see what we think.

Sam continues to escape, by fair means or foul. Chickens have lost their appeal for him and his efforts are now concentrated on the charms of the local bitch population, or the delights of Madame Cazin's milk bowl.

James's trips back and forth to London continue. And Geoffroy's trips back and forth to Gageac increase as he spends more time imposing a sense of order in and around the grounds of the château. A real friendship has grown between us. Each time he returns to Gageac from Antibes, he calls to see whether I'm free. As his boyfriend lives in Lyon he is on his own for the most part, and we often have supper together. There are certain parallels between us. He is a sociable but intensely private man and lives two lives, one professional in Antibes and the other here, where he was born.

I, too, have two lives; one with James and one without him. I am getting used to driving to and from the airport to deposit and pick him up, but not to him going away. The road from Gageac to Bordeaux, which I now know well, represents loneliness for me. It's a three-hour round trip and as I return in the car on my own, I listen intently to French radio to avoid dwelling on my thoughts.

Geoffroy's love of life and of his parents is apparent. His quiet discretion and genuine love of people create an instant rapport between us. Like his brother, Jean, he laughs as he talks, always animated and delighted at recounting tales of his travels, or of Gageac.

The church at Gageac, he has told me, was the important seat of an ancient archdeaconship in the eleventh century. It was named after the Virgin Mary, l'église de Notre Dame, who appeared to one of the residents in Gageac. As a result, it became the site of regular pilgrimages. This may explain the candlelit procession I witnessed when I first arrived.

By 1220 it had grown to cover fifty-three parishes. In the 1230s the bishopric was split into two dioceses and the seat of the archdeaconship was transferred to Flaugeac. At some point, the church was destroyed, then was rebuilt in the eighteenth century to become what it is now, a simple, country church.

With the separation of church and state in 1904, all the furniture and effects became the property of the state, much to the chagrin of the peope. Government officials were sent to make inventories of the interiors of all churches. The faithful of Gageac, armed with scythes and secateurs, gathered in a show of defiance outside their church. However, being peaceful people, after an hour of prayer they locked and barricaded the doors of the church and departed. When the doors were finally opened with an axe, the officials found a notice hung on one of the chandeliers: *Fermé au cause de décès: La Liberté est Morte. Défense d'Entrer Sans Permission du Public.*

James looks serious and tired. 'The recession is really biting in England. Loads of redundancies. It's tough.' He has returned a day early from his usual trip to London with the news that two of his consultancies may dry up. 'Doesn't seem so bad in France, but it's on its way here too.'

Chapter 4

'YOU COULD TRY ANOTHER LAB' SUGGESTS RICHARD GLUMLY TO James; 'they could be wrong.' James turns to me, urgency and panic in his voice. 'I need you to take a sample to another laboratory, the one on the Ste Foy road. It's on the left, called Tabouy.' He is looking alarmed and pale.

'What's the matter?' I ask. The matter is the wine, which is in the process of turning to vinegar. I want to ask how, why, what has made it suddenly all go wrong? And why am I going to the lab when I know nothing about wine or winemaking or analysis or samples? I start to voice my thoughts, but look at James's face and decide not to.

In a panic now myself, I get into the car with a small bottle of the red wine. '*Échantillon, échantillon,*' I repeat to myself; 'sample'. 'Here is a sample. *Voici un échantillon.* Please will you analyse it.' Oh God, why me? I don't know how to explain.

I arrive at the lab and get out of the car with the sample. My hands are sweating and my head is pounding. '*Voici un échantillon,*' I say to the stern-looking woman at the desk.

She looks at me and the sample and babbles at me in French.

I try to explain that the wine is turning to vinegar. Please would they analyse it?

*　　*　　*

The result comes two hours later by fax. James picks it up. The wine is, in fact, now vinegar. The whole of the red harvest is lost.

The effect on him is awesome to witness. He sits down on the chair nearby. Dazed and uncomprehending, he seems to shrink visibly in it. He stares through me. The only recourse is to ring the distillery and ask them to take it away.

'Can't we bottle it ourselves as vinegar?' I suggest, stupidly. 'Have you any idea how many bottles that would make?' says James. No, I haven't. I have no idea of how much red, or white or sweet wine, or vinegar, there is in the *chai*.

The weeks that follow are miserable. The weather is equally depressing, with unremitting rain replacing the summer heat and dust. The red grape harvest, of which James had such high hopes, is gone. He is desolate. He wanders in and out of the *chai* to look at the wine, returning only to sit down in the kitchen and stare into the middle distance. When he speaks, his voice is quiet and grave. A curtain of black gloom descends on the house and on us and I feel a presentiment of failure. The white, even to my uneducated palate, is pretty acid, which leaves us with only the sweet wine.

Another week in London does nothing to relieve his mood. Things are even worse there, and James's fear that he would lose two of his consultancies is confirmed. Back home, he visits the bank at Gardonne to discuss a loan for improving the *chai*, and returns pale, ill and defeated.

The bank has refused to grant him a loan, my first inkling that we have real financial problems. James says he will have to talk to Serge and trim down the proposed improvements to the *chai*.

'Why won't they lend the money?' I ask.

'The project is too big for the size of the property and we don't have enough collateral,' he says.

'And the purchase of the house and vineyard has taken all our capital,' he adds.

James's last statement hangs in the air, huge and heavy, then strikes me with a force that takes my breath away. How could I not have known that our move to France involved a serious risk and that the consequences could be disastrous? My mind is reeling. I look at him in disbelief and we stare at each other in silence.

A lorry from the distillery arrives to take away the red wine, and the following day an ambulance takes James to hospital. He has what seems to be a recurrence of the illness he had before we moved to France. With the help of Rosemarie, we explain to the doctors what we know of his last illness. The doctors don't agree with the diagnosis given in England, but don't know what is wrong with him either and there follows a week of tests in the hospital at Bergerac. From there, he is transferred to the much larger hospital at Bordeaux for further tests.

His sense of isolation and fear increases. He is ill, unable to express himself as well in French as he can in English; worried because he cannot get to London for meetings with his remaining clients; and unable to oversee the start of work on the new *chai*.

I make endless trips to and from Bordeaux to visit him, filled with dread at what the doctors might find. He in turn, finds it difficult to comprehend what is said to him by the doctors as he is too exhausted to concentrate.

Three weeks later, and he is home again, very weak and confined to bed. The feeling of impotence and desperation is compounded by worry about his state of health, both physical and mental. The doctors say he should get up. His illness, four weeks of lying in hospital beds and a deep depression have weakened him to such an extent that he simply cannot.

*　　*　　*

'*Bonjour, Madame. Vous n'avez pas reçu ma lettre?*' It is Bruno, our *oenologue*, here to sample the wines in the *chai*.

I try to explain that James is ill in bed, which is why I am standing here, feeling more than a little stupid and perfectly aware that I have absolutely nothing to contribute to a discussion about the remaining wine.

He asks for a glass and tastes the dry white, running it over his tongue and making a strange, gargling noise. He tells me the wine is a little acid and a little oxidised and that he will take another sample to the lab and call me.

Having no idea whether oxidation is good or bad, but assuming acid is not what we want and, all in all, preferring not to think about it at all, I am relieved to say goodbye to him.

'You can't possibly go to England!' I exclaim. James plans to go back on his own to recover and to find a job or more consultancies. He has been diagnosed as having systemic lupus, an autoimmune disease that varies widely from individual to individual. Its cause is unknown and the treatment varies according to the intensity of the symptoms. He has been told to minimise any sun exposure. Emotional stress can make the condition worse and flare-ups alternate with periods of few or no symptoms at all.

'One of the experts in the field practises in London,' he says. 'Besides, we simply don't have enough money to continue now that his consultancies have dried up.' The only option is to return to England where at least there's a chance of finding work. 'We can just sell up here,' I say. Vines and wines are of no consequence to me and, anyway, I don't want him to go back on his own. The idea of staying here without him is unthinkable.

'We can't do that,' he replies. 'Vineyards or houses in France don't just get sold overnight.' He is going to stay in England until he finds something. It won't be forever. And his staying here doing nothing won't help anything.

My mind is reeling. What will he do? What will I do? I really don't want to continue here without him. I don't know anything about vines and I can hardly speak French.

The decision was made. The journey from Gageac to Bordeaux on that cold winter's day was wretched. Turning to leave and waving goodbye to him was agonising, and returning to Gageac and the house only echoed the loneliness and sadness I felt.

A month passed in which I tried to come to terms with the reality of my situation. I spoke to him daily on the phone and he seemed to be coping, unlike me. I sank into utter despair.

If my life in France had been a steep learning curve to this point, it was nothing compared to what was in store for me. Learning to live on my own, learning how to make wine, getting serious because I had to and then because I wanted to and being a woman in very much a man's world were only some of the elements involved in what was to be the roller coaster existence of the next few years.

Starting with Boujema's announcement that he can no longer work his two days a week for us as he is now working full time for someone else. The initial feeling of abandonment, resentment and personal distress is replaced with the awful realisation that there is no one else but me to do the work of the vineyard. The arrangement with Richard is finished, James is in England and Boujema is gone. If I am to carry on here atall I will have to get to grips with the basics of winemaking and the work in the fields.

And I will have to start right now.

Chapter 5

TELEPHONES CONTINUE TO BE INSTRUMENTS OF TORTURE FOR ME.

'*Il faut soutirer la cuve.*' Bruno has telephoned. I have no idea what *soutirer* means, no idea how to do it and no vocabulary to ask for an explanation by telephone. I can only write down the word phonetically, then ring Richard to ask what it means and whether he can show me what to do.

Lesson one: *soutirer* is to rack. To rack is to draw off the wine from its leas. My introduction to the chai is with the racking of the dry white, as per Bruno's instructions. It requires pipes, a pump, a stepladder and willpower. Plus two large vats. I soon learn that nothing happens in the *chai* without them.

So begins my life as a winemaker. So also begins my life as a steeple-jack, climbing up ladders carrying heavy pipes, clambering on top of huge vats to place pipes in them. First attach two pipes to a pump – one from the full vat to the pump, the other from the pump to an empty vat. The first pipe must be immersed in wine from the top of the vat, so I need a five-metre stepladder. Climb it and place the pipe into the liquid, making sure it does not rest near the bottom where the leas sit. Then pump the clear juice into the above mentioned empty vat and stop as soon as it becomes cloudy.

To attach the pipes to the pump you need a grip of iron. To transport the pipe up the ladder to the top of the vat requires the shoulders of a bodybuilder and the muscles of Hercules to say nothing of nerves of steel at the sight of the drop opening up beneath. To place the pipe in the wine without disturbing the leas requires a delicate touch and to switch on the pump when everything is in place is impossible as you are at the top of a ladder and the switch at ground level.

So also begins my life as an ace tractor driver, hauling spraying machines, mowers, cutters and choppers.

'*Non, non! Ce n'est pas comme ça qu'il faut le faire!*' Gilles Cholet is standing before me. For the last hour or so I have been singularly ineffective, trying to work the drive shaft in order to '*broyer le bois*'. The dead wood that Boujema pulled from the vines and placed in the middle of the rows before his departure now has to be cut into tiny pieces, which is what '*broyer le bois*' is; mulching. This is done with the aid of a tractor and *girobroyeur*.

I have lurched, I have stalled and I feel utterly shaken so far. I am in third gear when I should be in first and in low throttle when I should be in full. Gilles explains all this patiently and I set off down a row, terrified at the noise and worried about how I will turn into the next row once I get to the bottom of this one. He runs along beside me, shouting instructions. As testimony to the power and volume of his voice, I can hear it even over the deafening noise of the tractor and *girobroyeur*.

Every so often, he shakes his head, orders me off the tractor and hops up to demonstrate the correct method. Then I get back on. 'Keep your foot off the clutch! Why have you got it on there in the first place? Full throttle! Go on!' And '*Voilà! C'est bon!*' when I eventually manoeuvre the tractor down the length of a row at full throttle and cut up the wood without swerving right or left.

My sense of relief is immense when, reaching the end of a row, I lift the cutter from the ground and cut the throttle, restoring relative silence and a sense of control. Then I labour to turn into the next row, and almost demolish the vines at the end of it.

'No!' screams Gilles as I lower the cutter, 'Lower it gently! You'll massacre it!'

I open the throttle to full and off I go again.

When I dismount from the tractor at the end of the day, my legs are shaking with the effort of depressing clutches, accelerators and choppers and my arms are trembling from the throttle and the shuddering wheel I have been directing. I feel strangely light-headed and shell-shocked but relieved to have achieved some sort of mastery over the machine and, more importantly, to have done some of the mulching.

The white wine is racked and the next hurdle is the laboratory. The CIVRB is the Interprofessional Confederation of the Wines of the Region of Bergerac, and in it is housed the laboratory. It is situated on the edge of the river at Bergerac. A huge wooden door of ancient oak with nails embedded into it, behind which are stairs, leads up to the laboratory wherein lie the *oenologues* and technicians, veritable gods and sages. This is where Bruno works.

The door is the first challenge. No matter how I try, I cannot open it. It looks like a latch, but doesn't lift. It also looks like a knob, but doesn't turn. I lift and turn at the same time which also doesn't work, then put down the sample I'm holding on the pavement to get a better grip of the handle. The bottle tips over and some of the wine runs out onto the ground, as a man appears with four or five samples in one of his hands, deftly positioned between thumb and fingers. With his free hand, he depresses the handle, for handle is what it is, and opens the door. 'After you,' he says.

I mount the stairs, feeling deeply insecure. A small counter is positioned at the top of the stairs, where three men are standing, deep in animated conversation. My heart sinks. They are huge and intimidating. Next to them is Bruno. Hopefully, I can just slip past them, deposit the *échantillon* on the counter and disappear before anybody notices that an interloper has been in the sanctum.

All conversation stops as I approach and all vocabulary I might have had vanishes with it. My mouth is dry, my eyes are fixed on the counter, my hand is clutching the sample. '*Bonjour.*'

Bruno steps forward. He introduces me to the assembled group.

I shake hands, give him the sample which is now only half full and try to remember that bloody word. '*Voilà l'échantillon,*' I mumble.

He thanks me. Did the racking go well and have I managed to do it without taking any leas? he asks, smiling.

I want to say that I did take some of the leas and that the racking didn't go at all well, but instead I say '*Oui. Voilà. Au revoir,*' and flee down the steps to the safety of the car. Thank God for *voilà*.

I sit in the car outside the lab. It is truly ridiculous that I am so intimidated by these people. Why? But I know perfectly well why. I know nothing of the business of wine and I can't speak French. I feel a fraud and I'm a foreigner. They all stop talking and gape at me because they can see that I know nothing. It's an insult to them and their business. And I'm a woman to boot.

I drive home, depressed and demoralised. I hardly know how to turn on the tractor, and the pipes and pumps in the *chai* terrify me. I taste the wine from the vat and don't know what I'm supposed to be looking for. I do a racking and haven't the faintest idea why I'm doing it. I understand less than half of what anybody is saying to me, if that, and every time I think I'm making progress in the language, I seem to regress. And as far as any technical language is concerned, I haven't even begun. And furthermore, I don't want to be here at all.

I miss James. I've stopped crying about it now but it all seems so utterly hopeless. I think of him back in England and worry about him, worry about money, worry about how I'm going to be able to continue here, and why I am even trying to. I know the answer to that too. It's because James has asked me to.

Bruno arrives with the results of the analysis. '*Il faut resoutirer.*' He looks at me for a moment, then smiles and explains why, half in English and half in French.

We taste the wine together and, patiently, he clarifies what we are looking for in the glass, asking at the same time what I think. He separates out the elements; the aromas (there weren't any of consequence and what there were smelt slightly off), the taste in your mouth (acid and tart) and the aftertaste (there wasn't much and what there was wasn't that nice, bitter and astringent). The racking will take away some of the bitterness in the aftertaste and perhaps bring back some of the aromas, but won't take away the tartness.

I explain miserably and half in sign language how ineffectively I did the last one.

'Take the pipe out when the wine level is below the bottom of the door of the vat, long before the wine becomes cloudy,' he says. 'Then descend, open the door and draw what's left from there at ground level. It's much simpler . . . See you at the lab.'

What he didn't say was that although the racking would make a little difference, it couldn't mask the underlying problem. The wine was awful, already yellowing from oxidation and with no aromas in the glass because there weren't any to start with.

I swap my *Au Courant* for *Le Vin en Dix Leçons* on the basis of priority; then I decide I need both, plus Harrap's and my *Locutions Vivantes* 'for all students of the living language who wish to broaden their minds and enrich their knowledge'. '*Quand le vin est tiré il faut le boire*' is one of its useful phrases – you cannot change horses in mid-stream. How appropriate. The white owl breathes heavily as I close the shutters of the bedroom window and fall into bed.

'*Salut.*' Michel, the *cantonnier* is outside, pruning the hedge around Eida's field as I open my shutters the following morning. He looks at my hedge. I look at my hedge too, a sad and untended hedge. It has grown too tall, and great gaps have appeared in it as a result. 'It needs a bit of attention' he says. 'How's everything going? I see you working in the vines. Hard work, isn't it?'

You can say that again.

'*Attache les lats,*' I read in my wine book. The *lat* is a thin piece of wood sticking out at the top of the vine. *Attache* means fastening the said thin piece of wood onto the bottom wire of the three wires in front of me. In order to do this, the *lat* must be passed under the wire once then attached at the end by means of a cord. Easy.

Vine number one looks gnarled and empty. In fact it looks dead, except that the *lat*, on close inspection, has eight or so buds, not yet open. I must not, under any circumstances, knock off any buds as they are all potential grape suppliers. I must not break the *lat* for the same reason.

My initial enthusiasm is dampened as it simply doesn't seem possible to bend the *lat* sufficiently to reach the bottom wire, let alone pass it underneath once before attaching it to the wire. The *lat* keeps springing back towards me each time I let go and it keeps slipping out of my grasp, even though I am holding on to it tightly but carefully, in between the buds. I try holding it at the end, just after the last bud. Even worse, it springs back and hits me sharply in the face. Bloody thing. I hold the end, force it down and — snap! I've broken it.

'*Bonjour.*' Michel the *cantonnier* has reappeared as if from nowhere and is standing beside me. '*Je vous montre,*' — I'll show you, he says, and he takes the *lat*, using two hands, one at each end. In an instant he has bent it into submission, under the wire and round, deftly tying it on to the wire and cutting the cord with his knife.

'You try,' he says, and he witnesses yet another failed attempt and another broken *lat*. 'You have to coax it,' he smiles, and shows me again.

He has slipped his right hand under the wire before taking the end of the *lat*. His left hand moves gradually and carefully along the *lat*, from the base to the centre. Miraculously missing all the buds that are waiting to be knocked off and coaxing the *lat* down to bend under the wire, then over, his right hand moves on to the right side of the wire with the *lat*. A quick twist of cord and it's done.

Two or three more failed attempts and I've got it. My sense of achievement is out of all proportion to the actual task. I am delighted and set off down the first row. Halfway up the third, I look up and see thousands of vines ahead of me, all with *lats* sticking up. Oh my God, how long will all this take?

I return home that evening to find that Michel has pruned my hedge.

It is February and cold. Very cold. I have noticed that if it is dry, the *lats* break easily. If it is damp, they pass under the wire with ease. Who would have thought I would welcome damp days? Who would have thought I'd be standing in a field amidst thousands of vines in sabots and a woolly hat doing this? Who would have thought that happiness is more tied-in *lats* than not tied-in *lats*?

'*Vous avez besoin d'un coup de main?*' Do you need a hand? Roger Merlos, with his huge cauliflower nose and stomach is stomping towards me through the vines, wearing his regulation blue overalls and thick coat. It's morning and his cold breath hangs in the air. I look up in surprise and incomprehension.

I can understand Michel after a fashion, but Roger's accent is unintelligible. He is to become my companion in the vines for that first year of tying up *lats*, much to the delight of his wife, who is glad to have him working again and out of the house.

We work in pairs, one row apiece but opposite each other, and quickly fall into a rhythm, working at the same speed. If it's a dry day, we skip some rows and move on to *lats* that have a different texture and elasticity. I quickly discover that white wine *lats* are for dry days, red for wet.

I discover, too, that afternoons don't work too well for Roger, the effects of his lunchtime break rendering him not only incomprehensible, but also *hors de combat*. The first afternoon is sufficient to demonstrate his inability to walk, talk or work. He stumbles over the wires at the end of the rows, his voice is slurred and he sways dangerously around the *lats*, holding aloft his small, sharp knife for cutting the string in an alarming fashion. The enigmatic smile on his face and vacant look in his eyes are telling. Gilles has already warned me. ' *Ooui!* He once set fire to most of the hillside of Gageac! *Bien sûr*, it was after lunch!'

We reach an agreement, reluctantly on his part, that mornings will work better. I have got used to his thick accent and listen as he recounts how he used to work for Monsieur Bichon who has fifty times more vines and *lats* to be tied in.

'This is nothing,' Roger tells me, gesticulating with his head at my vines and the *lats*. He's been tying in for fifty-two years and it's only ill-health that has stopped him. At the age of twelve he was sent to live on a vineyard, board and lodge as wages.

I am horrified.

'Yes', he continues. 'During the vendange, I picked all day and pressed all night. We didn't have your modern presses!'

I see mine as an antique and have been told as much.

'*Non, non,* we had the circular ones.'

I have no idea what these are.

They require a team of men, he tells me.

Or young, under-age children who have been sold into slavery I want to interject, but can't because I don't know the word for slavery or the word for under-age.

'We pushed the press handles round and round all night once the vendange was in', he continues. 'That's where I met Pepita who was working on the property. She was a bit younger than I was.'

My God, younger than twelve?

They decided to get married. I want to ask how old they were, where their parents were, how they could have been abandoned like that. Instead I say, how nice.

'Yes, she was beautiful, Pepita. Spanish, you know. Not used to living in houses.'

Any response I might give to the last comment is foreshortened as we arrive at the last *lat* of the *parcelle* of vines and Roger trips over a vine stump.

My proficiency grows as I work. Fewer buds fall, fewer *lats* are broken and, all in all, I get used to being outside all day; at least there's time to think. I think about vocabulary, about what Roger might be saying, although it's impossible to understand him this morning. He doesn't seem to mind that I don't reply and continues to mumble happily. I decide he's probably still inebriated from the night before, which would account for his slurred words and heavy step. He has told me that in his last job he would start the day with a *casse croûte*, a breaking of bread, with wine as an accompaniment, as everybody did in those days. It has obviously become a lifelong habit. Workers were paid in wine as well as money, he had told me. I surmise that there must be a fair percentage of alcoholics among the agricultural workers.

I think, too, about what Bruno said yesterday, that I have a good palate and should enrol for the wine tasting or *dégustation* course run by Jean Marc Dournel for winemakers. I instantly said no as it would be too intimidating, too technical and my French isn't nearly good enough. Bruce swept aside these objections and insisted. I could feel mounting anguish at yet another obstacle looming ahead.

There is no time for senseless self-pitying reflection on loneliness or lack of finances as total, healthy exhaustion overtakes everything but hunger at the end of the day. Food never tasted better, a bath was never so luxurious and sleep never so welcome. As I close my eyes each night, I see vines and *lats* for at least ten seconds before sleep overtakes me. The white owl's heavy breathing in the lime tree outside the window is lost in the deep, heavy luxury of sleep.

Jean de la Verrie stops his van and walks over to where I am tying in. '*Bonjour. Comment allez vous?*' he asks, shaking my hand.

'It's a lot of work,' I say, looking at the vines.

He laughs and agrees. 'If you need any help, let me know. *Au revoir, Madame.*' And he's gone.

Madame Cholet, Gilles Cholet's mother, passes through the vines every morning en route to the cemetery, a bucket over her arm. I'm not really sure what the bucket is for, but she carries it with her all the time. Her life is ruled by a strict timetable. She rises early and works in Gilles's vines from 5 am until 8. Her stroll to the cemetery and back is always at the same time.

Small and slim, she walks with grace, but also with a jaunty swing of her hips. Her house is in the middle of the vines between one of my *parcelles* and Gilles's. She tends her vegetable garden and her roses before lunch and in the afternoons sits in a chair under the tree in front of her house, reading the local newspaper, a cat on her lap. Next to her house is a dilapidated wrought iron cross leaning backwards from its stone pedestal. She mows around it regularly. Odd that the cross is so dilapidated yet its surroundings are so well maintained by her. She, too, wears the obligatory sabots.

She's off to tend the grave of her husband who was stolen away from her twenty years ago, she tells us. She looks at Roger, then at me with her intense, sharp eyes. They are pale blue, full of intelligence and force. There are those who never set foot in the cemetery once their families are buried, she continues. Except on All Souls of course, but then it's only to save face. Just because they buy a huge pot of chrysanthemums once a year and go to mass doesn't mean they're any better than those that don't. She doesn't need to go to mass to prove anything. Can she pick some *mâche* which is growing in my vines on the way back, she asks. *Mâche? Oui, oui,* I say. It is far too complicated to ask what it might be and anyway I am ashamed I don't know.

'*Mâche*, Madame.' announces Madame Cholet. She is on her return trip from the cemetery and with a knowing look, shows me what she has picked. She's going to make a salad of *mâche* and *ortie*, lamb's lettuce and nettles. Roger says something incomprehensible to her. She glances at him for a moment, then at me. She has just finished tying up – '*attachée*'- she tells us, in Gilles's vines. She picks up her bucket and with a wave, saunters off through the vines with her jaunty walk.

We are almost finished too and fast approaching the last *parcelle* of vines, which, Roger has informed me, is a merlot *parcelle*. He repeats the phrase again, as he stumbles over a vine stump. His voice is thicker than usual and I can guess why. He's celebrating on my behalf, he tells me. As he turns back to the vines, the faraway look in his eyes prevents him from spotting the wire in front of him. He lurches into it, staggers momentarily, then falls heavily as I leap to the side.

The vines shudder as the wire takes his weight. A worrying creaking sound comes from the stakes and tensed wire, then a series of *lats* that have just been tied in spring back up with force, buds flying off as the *lats* stand smartly to attention. Like a beetle on his back, he cannot get up. I try to help as he babbles away to himself, deep guttural sounds emitting from his mouth. '*Merde!*' I decipher amongst the torrent.

*   *   *

The nearest villages to Gageac are Saussignac and Gardonne. Gardonne has a petrol station, bakers, grocery and bank and is three kilometres away. It also has a hardware shop run by Roland Feytout. Tall, with a beard and a mass of black curls interspersed with silver, he has natural charm and boundless energy. He wears long, colourful patterned trousers and sneakers. As he talks, he jokes and laughs, which he does all the time, moving from foot to foot as he does so, arms flailing. He speaks excellent English that he likes to practise at every possible opportunity.

The shop is a cornucopia. Everything from a nail to a dinner service, a hosepipe to an industrial waste cleaning machine, is to be found there. Roland also makes and installs conservatories, cuts glass and keys and is a general supplier. What he hasn't got in his shop he can get within a day or so. The shop also seems to double as the local social club and general meeting place and is always full of people. All the artisans use it as well as the local population. He is married to Beatrice, who is blonde, talkative and just as friendly as he is. Serge the builder is a lifelong friend. They grew up together. Their wedding dates are the same, so they couldn't attend each other's weddings, but they celebrate their anniversaries together.

'*Salut,*' says Beatrice brightly as I step into the shop. It is full of people.

'*Salut!* Have some coffee,' says Roland, and smiles as I approach the counter.

There are people having coffee, people chattering at the counter, others inspecting goods; a hubbub of noise. No one appears to be in a hurry.

'How are the vines?' asks Beatrice above the noise. 'You know, I've never tried picking. I wouldn't mind trying it. If you need a hand during the vendange, let me know. If I can't help, maybe Roland can.'

*   *   *

78

Saussignac has one village shop and a *boulangerie*. It also has a large château, a church and a *notaire*. From the tiny hamlet of Monestier, on my way back from the small *parcelle* of vines we own there, the village appears invitingly on the horizon, a cluster of rooftops, with the tall steeple of the church and the imposing outline of the château rising up above the houses. The château is large and severe, with high, square towers at either end. In one of them live our Swedish friends Pia and Ekan and in the other my American friends, Joan and Fred Montanye, who run a cookery school there. They have been there for fifteen years, *émigrés* from California.

It is 7.28 on a cold and dark February evening and I am sitting in the car outside the CIVRB with a sinking feeling in the pit of my stomach and slight nausea. I can simply drive away, I tell myself. I don't have to do this. But I do have to do this. It's too late. Bruno has enrolled me for the wine-tasting course and there is nothing for it but to get out of the car and walk over to the large door.

Yes, I am the only woman here. Yes, I am deeply intimidated. Yes, I want the ground to open up and take me anywhere other than here.

'*Bonjour, Madame.*' Our first *oenologue* Jean Marc Dournel greets me and I am still intimidated. I saw him that one time with James, but have never spoken to him.

Tall, with a steady gaze and assured manner, he has a very good reputation, having created the client base at the CIVRB and modernised the laboratory there. He is considered one of the best *oenologues* in the area and is also known for being very exacting; not someone to suffer fools gladly.

I sit down at a desk. I'm back at school. In front of me is a file marked *Cours de dégustation*. Inside are two more folders; *La Dégustation des Vins* and *Les Principaux Constituants Chimiques du Vin*. I stare at them maniacally rather than look up and be confronted by the other people in the room, which is fast filling up.

'*Bonjour*. David Fourtout.' Young, dark-haired with bright eyes and a large smile, David Fourtout is standing in front of my desk, extending his hand. He's the son of a vineyard owner, he tells me, so has been brought up with vines and vineyards all his life. He has just been given a small parcelle of four hectares by his father to experiment on. Do I have vines? he asks. He wants to make something rather different from his father, something more concentrated, of higher quality. It's a good idea for him to follow this course because even though he knows how to taste, this is in-depth and thorough.

I can see by looking at the two folders in front of me that it is going to be just that.

Jean Marc is talking. No one else speaks. I can't understand what he is saying. He talks quickly and assuredly, but with no inflections in his voice. I peer at his mouth. Much better, the flow of words takes on a melody of sorts; I can follow some of it, but I can't look at the folders because I need to see his mouth to comprehend. 'The perception of sensations,' he is saying, 'is a physiological operation that requires a sensitivity and an integrity . . . it is an intellectual operation requiring concentration on the part of the individual.'

My concentration is at maximum. God, what is an *organoleptique* memory? Or an *entraînement réfléchi* or even a 'delicate phase' which is often *escamontée*?

From time to time, Jean Marc looks at me and blinks, then looks away. I am sitting at the front of the class, the better to see his mouth. He looks at me again, blinks then looks away. He moves up the room slightly so that he is more or less adjacent to me. He moves further towards the back of the room and I have lost the thread.

I learned later that he was so fazed by the way I was staring at him, he had to get away. Little did he know that he would have to suffer it for the entire course, and the one after that.

'. . . Volatile acids . . . *L'ensemble* of the acids of the acetic series . . . the more the volatile acids increase, the more the wine is altered. The state of cleanliness in the *chai*, the technical qualities of a good winemaker, good storing conditions.' He is back in front of me and I have picked up the thread again.

We move on quickly from volatile acids to fermenting sugars, *ethylique* alcohol and glycerol, dirty tastes in the wine, bitterness and astringency and odorous substances and my head hurts a lot. 'That's all for today. We'll discuss azotes, pectins and mucilages on Friday evening.'

I rush home to the dictionary. 'Mucilages'; 'gum'; *'escamontée'*; 'vanish'; *'entraînement'*; 'training'; I pore over the files. 'Our appreciation of wines depends on our sensory impressions; vision, olfaction, gustation and tactile sensitivity,' I read. 'With experience, some synthesis of them all enters into our judgement. Our sensory appreciation of wines is due in the most part to the olfactive sense, which is located in the upper part of the nose.' The sense of smell, but a bit more than that.

I move on to the visual and gustative sensations, the aspect of the wine, the colour, nuance, effervescence, taste, touch and the interrelation of all of them, and my mind is swimming with yet another bank of newly ingested vocabulary. It is deeply interesting and I know that I want to know more. Why didn't I think of all this and learn the language before? But I did make it through the first lesson and, losing sight of Jean Marc's mouth notwithstanding, it wasn't as demoralising as I thought it would be.

Lesson two introduces the azotes, pectins and mucilages and the gustative vocabulary – and then to a tasting. In front of each of us is a glass, a piece of bread, a paper napkin and a tasting form. We are tasting dry whites. Jean Marc is explaining the separate elements. As he points them out, so they become evident and it is extraordinary to me that what I have taken for granted and never thought about when drinking a glass of wine should be so interesting. He talks of equilibrium, balance in the wine, harmony and fruit. He talks of acidity as a quality, giving freshness. He talks of the aromas, their intensity or otherwise, the sweetness or acidity, the tannins and the alcohol; all the separate components that make up a simple glass of wine.

He asks what we think of what we are drinking. He explains the taste zones of the tongue, and the subtlety of our olfactive sense, the complexity of the whole, the viscosity and the length in the mouth, where aromas and flavours combine and linger to a great finish. It is a revelation to me. The sensory quality of the wine in the glass, he is saying, is what is important. It is a very personal judgement that requires, above all, concentration. Concentration is not what the class is doing at this moment. It is a hubbub of noise as people discuss the separate elements they each find in their glass of wine.

'Comment c'était, la deuxième session?' Bruno asks when I see him a few days later. *'Fantastique!'* I reply.

I speak to James frequently on the phone. He has found some work, though nothing much. He's not coming back for a while, he says, as he must be available in England in case something else turns up. He's feeling slightly better, but fatigue is always with him and he hasn't the energy to do much. He's also seen the specialist in London, who hasn't told him anything different from what he was told here in Bordeaux.

'When are you going to come back and live here?' I ask.

I worry for him from a distance and long for us to live a normal life, a daily, normal life together.

I speak to Chantal, who he has been staying with. 'He's fine, Mum. Don't worry' she says.

*   *   *

Spring has arrived. Light streams in through the windows, clear and bright; rays of sun, bursting with energy and promise. The vines still have a dead, bare appearance but on closer inspection, the tiny buds I so carefully preserved or not when I tied in my *lats* have opened out into clusters of minute leaves. They are very beautiful, delicate and pale. Perfectly formed, first one leaf unfurls, followed quickly by two more, then an *épanouissement*, a blossoming. Looking out across the vines, what seemed like brown, gnarled emptiness only a week or so ago is now enhanced by a soft, green hue. This is replicated at ground level, both between and under the vines. I am about to discover two things: *épamprage*, cutting off the shoots that have grown from the base of the vine, and mowing at industrial levels.

Mowing means the trusty tractor and *girobroyeur*. I thought I knew how to use it from the *broyer le bois*, but, as if none of that had happened, I am once again at a loss to get the drive shaft to work. Through the vines I see Gilles advancing towards me, shaking his head, ready to give me a blast from his vocal chords.

Different job, different speed he tells me. And off I go, drive shaft on, throttle full, second gear this time and close to the right-hand row of vines and not in the middle. I must go up and down each row twice to cut the grass. And not miss the growth in the middle! And keep a straight line! And not damage the vines! And lift the mower just before I turn, *not like that!*

Epamprage is a different animal altogether. The beautiful, soft leaves that now crown the vines are also crowding the base, which, I am told, is definitely not desirable. They look perfectly harmless to me, but Gilles insists they will take the goodness from the grapes. 'You must cut them off now while they're young and fresh, it'll be much more difficult later on. And you'd better get a move on as you'll need to have removed them all before I teach you how to weed under the vines, and time's running out. And while we're about it, there's the crowns of the vines to be pinched out too!' And with a wave of his hand, he sets off back towards his house.

Consider the repetitive movements of bending down to the base of the vine, cutting out the shoots, not knocking them off, straightening up for a second, then back down to the next. Gilles had told me that for every one I knock off instead of cutting, five will sprout in their place. Multiply this by four-and-a-half hectares.

Michel once again appears from nowhere when I'm working in the vines, and with a '*Salut*', sets to and helps. As ever, he stays an hour or so, working calmly and speedily, an invisible helper, working discreetly and effectively. He is racing up and down rows leaving perfectly groomed, shoot-free vines. I keep finding an extra shoot or two lurking the other side of the base just when I think I've finished one vine. And what are these minute snails doing on the trunk? They are making their way up to the juicy leaves at the top is what they're doing.

'There's nothing you can do about them,' says Gilles when I show him. 'Yup, every year it's the same, they're always on the young vines; an invasion. They'll have chomped their way through most of your vines on this facing slope. There's no point spraying against them; might as well pour money down the drain.'

I start maniacally pulling off minute snails. Their shells break as I tear them from the leaves and my fingers hurt as I wrest them from the abrasive trunks.

'*J'exagère un peu*,' Gilles laughs. 'Of course they won't eat your vines, only a few leaves. And you're going to have more of those than you want anyway.'

Mowing and epamprage have a spectacular effect on the vineyard. Everything looks neat and tidy. Except the jungle growing at the feet of the vines.

There is no time to lose, says Gilles. I must weed systemically before the shoots grow again.

You mean they're going to grow again?

And he says he hopes I've been diligent in my work because if I've left a stray shoot, the weed killer will enter the vine and kill it off entirely.

I imagine four-and-a-half hectares of vines, each vine with a shoot or two left at the base, into which the weed killer will penetrate. My vines will be dead vines. And indeed, as I bend down to peer at the vine nearest me, there is a shoot. I hastily remove it and set off down the row.

'Come back!' shouts Gilles, 'If there are any left and you take them off now, it will be even worse — an open wound on the vine for the weed killer to enter more easily!' The vine will need a whole day to recover from any new cutting.

I can't possibly spray the weed killer, I tell him. I'm sure I haven't been diligent enough. Just two days ago, when I was flagging, I know I must have left some shoots and I didn't always look back at each vine as I worked up the rows. And there was one row that seemed to have almost no shoots and it did sort of lull me into a false sense of security.

Then I see what Gilles has now attached to the back of the tractor. It has a large reservoir, behind which is a long bar, extendable at both ends, with nozzles positioned along it. Festooned among the bars and nozzles and reservoir are countless blue plastic pipes. The bar will be extended according to the width of each row of vines, Gilles tells me. And lowered or lifted according to the range of the spray and the height of the vines. I am already confused.

My *parcelles* of vines do not have evenly spaced rows; some are narrow, some medium width and a few are very wide . The positioning and measuring for each row is vital, he says. The nozzle at each end of the bar must be properly positioned so that the spray reaches under the vine and not onto any leaves. We will close off all nozzles other than one at either end of the bar. As the tractor passes down the row, it will spray weed killer underneath the vines on either side, two rows' worth, in fact. If the bar is too high, the spray will hit the leaves and if it's too low it won't cover the necessary area. If I drive too slowly, I won't have enough force for the spray. And if I drive too fast I won't spread enough weed killer to do the job. If I don't drive in a straight line, I'll hook vines on the bar and the spray will squirt upwards onto the vines. I look at it and at them and I know I can't weed kill.

'Yes you can,' Gilles insists. He has fixed a set of three taps to the front of the reservoir, one for the right hand side, the other for the left – and warns me not to touch the middle one. They each have thin blue tubes leading to the bar. He positions the tractor at the top of the row, lowers the bar almost to the ground, then dismounts to extend the bars on either side to within a foot or so of either row. There is much standing back to inspect, with adjustments to one or other side. Then he unscrews each nozzle, taking out of them a small black disc with a hole in the middle. He blows through it, holds it up to the light, blows again, then replaces everything, wraps the spaghetti of blue tubes somewhere between the reservoir and the tractor, then tests the spray.

It looks like a piece of medieval war equipment. He opens the throttle and I leap back as fine spray shoots out of the two nozzles on either side. The spray reaches under the vine. A last adjustment from Gilles and he sets off down the row on the tractor. It is awesome to witness.

'Stop!' I shout. 'You've just sprayed a pampre!'

'It's minute', he shouts, 'and your eyes aren't very good 'cos I've sprayed at least forty!'

'Okay, get on,' he says, when he reaches the end of the row. 'Now make sure you keep in a straight line, stay at that speed and don't lift the bar by accident.' The nozzles look dangerously close to the vines and the spray looks too forceful.

My left foot is firmly pressed onto the clutch and the throttle is on full. Red tap for the right, blue tap for the left, don't touch the middle one – dare I go? I dare. Legs trembling, foot hovering over the clutch and hands clasped on the wheel, I set off up the row with a dry mouth and rising nausea. I can't look back as I need to concentrate on keeping a straight line.

Gilles is running beside me, 'More throttle!' he shouts, 'and take your foot away from the clutch!'

Lesson three of the *dégustation* course, and we move on to fixed acids in the grapes, glucose and fructose and the variant levels, *carbonique* gases, *anthocyanes*, flavones, condensed tannins and pips, skins and stalks. There is no point in bringing a dictionary with me as I haven't time to look up the words before a battery of new ones arrive. *Succinique* acids?

And so to the tasting. We are tasting from a specific area tonight; white and red burgundies. The bouquets are astonishing: the almonds, quince and lime of the whites; the raspberry, cherry and pepper of the reds. Am I finding these aromas only because Jean Marc is describing them? No, they really are there.

'What about the texture?' Jean Marc is asking. 'And the length?'

The vines are flowering. It is quite extraordinary. From the fresh, green leaves of a few weeks ago have burst stamens with tiny, delicately scented flowers. The flowers are hermaphrodites, Gilles has told me, and will produce the berries that will become grapes. 'So pray for good weather otherwise you'll have *coulure*,' he says. *Coulure?* 'And make sure your insecticide is delivered by tomorrow because you need to spray.'

The insecticide spraying machine is a monster. It is blue, with five huge trumpet-shaped funnels on either side, each attached to the obligatory blue plastic pipes that lead to the obligatory taps behind the tractor seat. It is truly terrifying.

I am following instructions and mixing buckets of blue powder with water, using a large broom handle. I have been ordered not to leave any lumps at the bottom. The tractor is idling next to me and the spraying machine is also humming quietly. Gilles and I tip buckets of the blue stuff into its reservoir while water from a hose dilutes and mixes it all up. 'Okay!' he shouts. I turn off the hose and he turns up the throttle. A deafening noise fills the air. Gilles switches on both taps and the blue liquid bursts out of the ten funnels. It sends spray in either direction with such force and noise that my first instinct is to dive for cover.

No such luck. I am on the tractor and heading for the vines, a by now familiar feeling in the pit of my stomach. I haven't quite recovered from the weed killing experience; indeed I still have the occasional nightmare about row ten when I hooked two vines with the end of the bar. The spaghetti of blue pipes tumbled down, the spray upended itself and the more I tried to extricate the bar, the further it lurched into the adjacent row, taking with it a vine. Gilles had gone home, having satisfied himself that things were under control. Michel had appeared from nowhere again, like some guardian angel, with the words: *'Ce n'est pas facile, tu es courageuse.'* Courage had nothing to do with it, fear was in the driving seat here. In two minutes he had righted the spray, replaced the pipes and extricated the bar.

The insecticide is loaded, the tractor is positioned, the taps are on and terror is mounting as I open the throttle.

Clouds of blue spray go everywhere. I am racing down the rows, faster than in any other task to date. I must keep the speed up in order to spray the right amount on the leaves. I must drive in a straight line. I must lean back, ready to stretch my arm to maximum to the taps behind me to switch off the spray at the end of the row. At the same time I must shut off the throttle and ram my foot down on the clutch to turn. I must also avoid the vine at the end of the row and position myself properly at the beginning of the next and check to make sure I haven't missed spraying the last two vines in doing so. Then it is taps on, throttle at maximum, and down the next row.

I am exhausted. I am covered in blue spray. My legs tremble as I walk back towards the house. My ears resound with silence after a day of supersonic sound, but I have a sense of achievement.

The *dégustation* course is finished and I am enrolled on the *Cours de Perfectionnement* on the basis that I have a good palate. I'm flattered and apprehensive in equal measures and relieved there will be a break in between.

But there is no break working in the vines. They are growing at an alarming speed. Where there were four or five delicate leaves from each bud, there are now long, thick branches, each with a canopy of leaves and tiny bunches of embryo grapes. They grow upwards and outwards initially, but are now cascading down to ground level. Gilles tells me 'You've got to get all those wires lifted before the next spray, otherwise the tractor won't get through.' Oh God, when is the next spray, I want to know, and what wires?

The next spray is in eight days or so and the wires, two for each row, are on the ground. They have been cunningly hidden by the weeds, except for their ends, which are attached to a stake at either end of the row. They need to be raised to shoulder height on the stakes positioned at one-metre intervals along each row.

Gilles picks up the wire from the ground with both hands, wrenches it out from the weeds and pulls it towards him, then lifts it up above his shoulders and hooks it onto a nail in the stake. As he does, the wire gathers five or six of the huge branches that have dropped to ground level and lifts them up to the trellis. He works steadily, taking the wire in his hands and stepping back with it. Outwards, upwards and another swathe of vine branches is brought into line when he hooks the wire onto the nail. I work on the other side, lift out, up and onto the nail. We reach the end of a row and look back to see the difference. It is very different. Where there was a bedraggled and sprawling mass, there are neat rows of upstanding vines.

'Okay. *Bon travail*,' he says, and is gone.

I continue. One small row of vines of one hundred metres needs to be worked four times – once down one side (out, up and on), then up the other side, then down each side again to tuck up the stray branches missed by the wires. Multiply this by four-and-a-half hectares, then repeat the process with crochet links and clip them onto the two wires, three per metre.

Roger comes to help occasionally, always in the mornings. Either he walks with a heavy foot, face concentrated in serious contemplation of the job in hand, or with a mincing gait accompanied by a seraphic look in his eyes and an enigmatic smile.

I am beginning to develop arm and leg muscles that I didn't have before. I am also beginning to know my vines, albeit in the most superficial way. The leaves are large and beautiful. They are healthy and green and the stamens have grown. They are now identifiable as embryonic bunches. The berries are hard, small and green, the bunches growing up and outwards, rather than downwards as I would have imagined.

I look up *coulure* after Gilles's warning and discover that it is one of many horribly undesirable conditions in the vines whereby the flowering is unsuccessful. No flowering, no grapes. Pollination is successful in warm, dry, stable conditions, but not in cold and rain as the flowers can't form properly and therefore don't yield grapes.

Every evening I inspect the flowers. I'm not sure what I am looking for. Indeed, you have to look pretty closely to detect a flower at all. They are tiny petalled, with the merest dot of pollen crowning the hearts. And although I find some, how do I know how many aren't even forming? I could be missing hundreds of them, and not know.

The man that supplied me with the blue stuff for the vines also gave me a horror book entitled *Maladies de la Vigne*, which promised to: 'aid in the identification of virus diseases of the grapevine, with one hundred and eighty-six photographs in colour of prime symptoms with other causes, such as bacteria, fungi, mineral deficiencies, insects and mites'. In it, I found that vines are susceptible to black rot, red rot, mildew, oidium, grape worm, *cidadelles*, *acariose*, and *botrytis*, to name but a few.

For the moment, my supplier tells me, I am spraying *bouille bordelaise*; blue stuff, otherwise known as copper sulphate, against the oidium. A gruesome illustration shows a vine afflicted with oidium. Its leaves are shrivelled and the grapes are split and imploded, looking unhealthy and rubbery. Tomorrow, I will be adding to the *bouille bordelaise* a treatment for mildew. I look up mildew, along with the accompanying illustration. I'm not sure which is worse, the worry of the maladies I might find or the prospect of getting on the tractor again with the hideous spraying machine.

The vines continue to grow. After only two days, the branches I had brought into order by lifting the wires seem to have grown by at least a foot. They are starting to keel over, in spite of the wires. And they are also growing outwards. It doesn't look possible to drive through the vines without hitting them on either side.

'*Bien sûr c'est le cas*,' says Gilles, 'you must *épointer* beforehand,' trim the vines. 'You don't want to throw money away by spraying leaves you're going to cut off. And you need to create some space for the spray from the top two funnels to reach the row beyond the one you're in so you can economise on spray and time. You can't spray over the vines with them at that height.'

If I had thought that the spraying machine was the ultimate in instruments of torture, it was because I hadn't seen the trimmer, the *épointeuse,* at close quarters. Sitting adjacent to the driving seat on the tractor, it is a system of lethal, sharply honed rotating blades attached to bars that straddle the vines; one on the nearside, a short one over the top and a third one down the outside. It must be positioned precisely to cut the leaves around the sides of the canopy and over the top. Ten double sets of blades run along each side, with six along the top.

I am truly horrified. Gilles positions the machine at the top of a row and explains that the height and width must be adjusted, the height to at least six inches above the highest stake, so that the blade doesn't hit it or the wire on top of it. If it does, the wire will be severed. It must also be wide enough not to cut off the grapes, but narrow enough to remove the excess leaves.

I am sitting on the tractor, engine running, with forty-six blades whirring at my side and over my head.

It is absolutely essential to drive in a straight line all the way along the row, Gilles tells me. 'It is carefully positioned. If you move out slightly, you'll cut off the grapes on the other side and possibly a wire or two, which could hit you.'

This is sheer madness, I tell myself.

'Keep in a straight line, keep the throttle at the same speed.' I set off. 'And be careful of the branches falling on your head! They're heavy!'

He is not exaggerating. The blades spin and as I look up at them at the beginning of the row, a huge branch falls on my head, followed by four more. I hunch up and fix my gaze on the distance and the end of the row while branches and leaves rain on my head and shoulders. The aim is survival; to get to the bottom of the row without killing myself. I no longer care if the grapes on the other side are being cut off. I continue in a completely straight line to avoid garrotting myself with one of the wires that will almost certainly be severed if I deviate by even one centimetre. The whirring of the blades cutting off the branches with speed and violence is terrifying. The severed branches hit my head, shoulders and back with shocking force.

I reach the end of the row, my eyes smarting with leaf particles and tears, and my legs trembling with the effort of keeping the same speed and direction. Branches hang from the tractor, the machine and me. Gilles is waiting at the end of the row, having run along ahead at a safe distance. '*Très bien fait,*' he shouts, '*très bien fait!*'

'You can relax a bit now,' he says. 'The trimming is done and the second *traitement*. Cutting off the shoots shouldn't be too bad at the moment. And it's time you passed the mower again. The grass is growing.'

The *Cours de Perfectionnement* has begun and Jean Marc is almost at ease with my staring presence. David Fourtout is in the class, which is much reduced in number. We listen, concentrate and taste. We taste solutions of components of wine in concentrated doses, and in diluted doses. We have triangular tastings, where three samples are presented, two of which are identical. We taste wine from every region of France.

The church at Gageac is used every other Sunday. The bells call the *Gageaçois* to mass half an hour before the service. I wonder where all the people come from. There can't possibly be that many houses in Gageac. I learn later that there is a shortage of priests, and our priest shares himself around four parishes. Parishioners from the other three trek up the hill to Gageac and mass every other week.

I love the church. I love its simplicity and I love its history, as recounted by Geoffroy. Before and after the service, people congregate outside the church. I can see them from my kitchen window and hear their voices if the window is open. Mr Cazin rings the bell both before and after the service, and is answered by the howling of Sam and Luke. I'm not sure why they do this, but it's patently a joyous experience. It is a definite howling, heads raised to the skies and voices giving full vent to the job in hand. If the wind is in the right direction, they can even howl to the bells of Gardonne, almost three kilometres away, and Saussignac too, the same distance, but up the hill instead of down.

Today the bells ring, the dogs howl and the *Gageaçois* arrive for Mass. It's a beautiful day with clear skies. The church bathed in morning sun. The organ sounds, the people enter, the door closes and Mass is in progress. The dogs this morning continue to howl and rush up and down the garden frantically, even after the bells stop ringing. The cats, too, are restless and Eida is rushing around her field, expending more energy than I've ever seen her do.

As I watch her, a young colt appears from nowhere. He gallops down the road past the church, then turns sharp left at the hedge opposite the house. He screeches to a halt in front of Eida who, by now, has galloped to her habitual spot in the field exactly opposite the house. There is a moment of silence as they nuzzle each other, then the colt is gone, off up the road towards the château. Eida gazes at me, a by now familiar quizzical look on her face.

People are leaving the church and congregating again on the grass as Monsieur Cazin approaches the house with our cat Edward under his arm. 'Madame, Edward came to Mass. He wouldn't leave . . .' As I listen politely I see, over Monsieur Cazin's shoulder, Eida, whose quizzical gaze has changed into one of force and determination. She steps back, lowers her head, kicks her back legs into the air, neighs and leaps over the hedge.

The effect is electric. Sam and Luke scoot past me and out of the door into the road while Edward leaps out of Mr Cazin's arms. Eida, by now intent on freedom and pursuit of the colt, careers along the road past the church and up through the vines to the ridge, scattering the congregation on her way. Seconds later, twenty sheep, who live behind Jean de la Verrie's atelier on the ridge, are rushing down through the vines.

All sense of Sunday Mass decorum evaporates. Jean de la Verrie rushes over in search of a rope to catch Eida, shouting directions to his worker at the same time as his sons run off into the vines in pursuit of sheep and horses. '*C'est probablement à cause de la lune. Une drôle de journée,*' remarks Monsieur Cazin quietly.

I am becoming rather good at attaching the various appendages to the back of the tractor. I can now attach the *girobroyeur*, the mower and the spraying machine. And furthermore, I can drive in a straight line and even turn reasonably well at the end of each row. We are now through the frenetic period when everything happens in the vines at the same time, and I slip into a routine of spraying every twelve days or so and mowing every two weeks. I have even cut off the shoots again, with the help of Michel.

Spraying the vines now with confidence, I hurtle down one of the rows and, turning on uneven ground with a gentle uphill slope, the spraying machine suddenly drops off the back. The noise is deafening; a high pitched screeching with blue spray emitting from every funnel and covering the ground.

Gilles is running towards me, having heard it from his house, as is another neighbour. We stop the machine and Gilles inspects the damage. '*Il faut aller voir Monsieur Bonny,*' he says.

Monsieur Bonny is essential to my life, indeed, to all vineyard owners and farmers in the area. There are two Monsieur Bonnys, each with their own atelier, in which they repair tractors old and new, along with any of the myriad agricultural machines that exist. Every nut, bolt and screw known to man is housed in one or other of the ateliers and a workforce of five to six people are on hand in each atelier to do repairs. They also build those ultimate instruments of torture, the trimmers, and sell second-hand equipment, as well as new. And they rarely suffer from quiet periods.

Today is no exception. There are three tractors with their engine parts spread over the floor, plus several farmers buying nuts, bolts or screws, awaiting service and talking among themselves. Some are looking at the second-hand equipment displayed on the lawn in front of the shop. There are sprayers and trailers, along with presses and trimmers. Gilles has helped reattach my sprayer to the back of the tractor and I have driven it over with Gilles following in his small van. Fewer than half of my vines have been sprayed and I'm desperate to get it finished before the rain, which has been forecast for tomorrow.

'Mmm . . .' Monsieur Bonny inspects the damage, stroking his chin with a black, oily thumb and forefinger. '*Je ne sais pas* . . . I don't know whether I've got the piece you need in stock.' He continues, 'I don't know if I have the time . . .' There is silence, an agonising silence. 'But maybe we can improvise . . .'

I stand, looking from Gilles to Monsieur Bonny as they discuss the pros and cons of an idea Monsieur Bonny has. 'Yes, that could work.'

My skin is starting to itch and as I look down at my arm, I see I'm covered in the fine blue spray of *bouille bordelaise*. In the time it takes to find the tap outside, wash as much of it off as I can and return, Monsieur Bonny and his workers have dismantled the machine and are hard at work.

'*Tu as de la chance!*' says Gilles, smiling and touching his nose. 'It'll be ready in half an hour.'

It is hot. As I drive back along the road from Monestier after spraying the *parcelle* of vines there, the landscape strikes me again with its beauty. The roadside is fringed with ferns and wild flowers and the undulating slopes to right and left are swathed in the gentle, green curves of velvet that are the vines. Above, the sky is cloudless.

A mantle of swirling white cloud suddenly appears and with it a hot, oppressive wind. The sky changes from clear blue to a brooding grey and pink. A deep, leaden sky hangs low and menacing over the vines; a storm is coming. The air is heavy and ominous as I drive along the road, intent on reaching home before it breaks.

Too late. A crack, and jagged lightning soars across the sky in front of me, followed almost immediately by a roll of thunder, terrifying in its intensity. The thunder echoes and rolls around the sky, in front, behind and down the slopes towards Gardonne, tempestuous and frightening. Another jagged, white light pierces the sky, followed by a booming crescendo of sound as rain falls suddenly, a curtain of water. Bouncing off the road, teeming and violent, it pelts the engine of the tractor and stings my head and body with its severity.

Panic rises in me as I try to gauge whether it will be safer to get off the tractor and take cover or carry on. Deciding that the rubber tyres will protect me, and with lightning coursing across the sky, I reach the brow of the hill over Gageac and descend towards the château, dramatic and beautiful against the dark sky. Thunder rolls around the skies, backwards and forwards as I drive down past the mulberry trees lining the road and arrive at the lime tree next to the house. Then a sudden silence as the storm abates and the rain stops as quickly as it started.

The man who gave me the *maladie* book has also given me a calendar of spraying times, along with the appropriate spray for the periods of mildew, odium, black rot and the like. In permanent fear of them all, I scan the leaves as I spray or mow. Interestingly, speeds that seemed record breaking when Gilles first set me on the tractor now seem gentle, almost leisurely circuits up and down the vines.

The bunches on the vines are now the size one would expect, but still green, hard bullets instead of soft swollen fruit. There is something very beautiful about them, even at this stage. They are clean and fresh, sleek and slightly lustrous. The leaves are large and equally beautiful, giving goodness and sugar to the grapes, a symbiotic relationship.

Geoffroy is moving from the château and into one of the small houses he owns and rents out in summer. It is opposite his brother's *chai* and the château and almost opposite the derelict house. He's decided to have some independence from his parents in the château, while being near enough to be at hand for problems. But he'll keep his apartment in Antibes, he says, as all his friends are there.

James is back for a week. It's such a pleasure to see him and to have him home. I show him the vines and the work that's been done in them, we talk about Gilles and Michel. James hasn't made much progress in finding a job or consultancies, and says he still feels tired and unable to cope. This I already know as we talk to each other frequently by phone.

Odile rings to invite us to supper the following evening, but we decide against it as James doesn't feel up to an evening of French and having to make an effort when he needs rest; he feels the same when Geoffroy invites us to his new home later in the week. The week passes and we are both miserable. Each time James gets up and decides to do something to help me, he finds it too taxing, which makes him depressed. We say goodbye again at the end of his stay with sadness.

He didn't enjoy his week and he wasn't that interested in the vines or much else. An overwhelming wave of black despair and loneliness engulfs me. The dream he had is fading fast, I can see that clearly.

I listen to the messages on my answerphone, hoping there's one from James to say he's arrived in England safely.

'Hi, Mum!' It's John, ringing from Thailand. 'Just wondered how you are, on your own over there. I'm thinking of you.'

Far from lifting my spirits, it sharpens my sudden despair. Andrew has a job teaching in Ecuador and Sophie is already in Boston. Emma is busy working and living in Aubusson, which, although only four hours away, could be the other end of the world. James and Chantal are in England and I'm in Gageac.

Among the winegrowers of Gageac there is a sense of unity and comradeship. Monsieur Cazin mutters about mildew and oidium and asks whether I have any. He has a touch in one of his *parcelles* on the hill. He can't understand why, as he sprays all his *parcelles* in the same way. Gilles shouts and raves about the grape worm he has on some of his leaves and Jean de la Verrie talks of his *cicadelles* and the fruit set on the grapes. I feel humbled by their assumption that I'm in any way an equal at spraying and mowing, or anything else, but also honoured to be counted among them. It gives me a sense of belonging that compensates for the deep loneliness I feel.

My spraying machine has broken down again and it's Sunday. Monsieur Bonny is closed for the week and Gilles is away. I am desperate. There is mildew on some of the leaves and I am in a race against time; the spraying can't wait until tomorrow. Standing beside the tractor in the vines with one of the funnels of the sprayer in my hand and a feeling of total impotence and despair, I look up towards the ridge and see Jean de la Verrie's tractor moving up and down his rows of vines. He, too, is working on a Sunday, no doubt with the same problem as me. He continues spraying for two more rows, then stops, dismounts and walks towards his van. His form is silhouetted against the blue sky as I turn back towards the house in dejection.

'*Bonjour. Vous avez un problème, je suppose?*' Jean de la Verrie steps out of his van and walks towards me, hand extended. He has noticed that I have stopped spraying. Can he help? He walks over to the tractor and looks at the sprayer, then at me. It is not going to be easy to fix, he says. And Monsieur Bonny is closed. '*Vous aussi.* You also have some mildew,' he continues. I nod, as we look at one of the vines with mildew. He walks further along the row and identifies some more. 'Yes, you have. What are you spraying with? The same as me, I guess.' Then, '*Vous ne pouvez jamais rester tranquille avec des vignes.*'

He gets into his van and drives back up to the ridge and his tractor. He climbs on, starts it up, then slowly drives down the hill towards me and comes to a halt in my vines. '*Tenez.* You know how to drive it, I know. I'll spray mine tomorrow.'

Breaking his own rhythm of work in a crisis to help me, when time was of the essence, was an act of extraordinary generosity. There would be many more instances of it, too numerous to mention.

Chapter 6

IT IS JUNE AND GAGEAC IS IN FULL SPLENDOUR. ALONG THE SIDE OF the house is a wisteria that runs up into the lime tree. It's in full blossom, dazzling pale lilac against the pink of the two tamarisk trees behind it. Their perfumes mingle, pepper from the tamarisks and heady musk from the wisteria. All in all, the small courtyard is now looking rather lovely. Michel arrived one morning, cut back the hedge again and sprayed the weeds. He also planted two rose cuttings from his garden, both of which have taken and are now blossoming.

The grass outside the church is mown and the mulberry trees along one side of Eida's hedge are covered in leaves. The wisteria in the courtyard of the château makes mine pale into insignificance. It is truly magnificent and, as if in recognition of this, the gates are permanently open, the better to display its splendour. Madame de la Verrie tells me her courtyard is a sun trap. When her children were small, she baked meringues on the steps of the château in summer.

When I recount the tale later to Gilles and Michel, Gilles says: 'My courtyard is so hot you can't stand outside at all. You hardly need to light a barbecue!' We are standing by the kitchen door, opposite Eida's field. She is standing at the hedge, tossing her head every so often to shake off the flies that land on her and listening to Gilles as Michel rubs his nose and nods.

Gilles continues: 'And I can tell that we're in for a hot summer, a veritable *canicule*.' (A quick dip into the dictionary tells me that *canicule* is a heatwave.)

Discussions follow on summers he and Gilles have known, *canicules* that have surpassed all others in intensity and resulting vendanges. Michel says if it's hot and dry in August, the harvest will be good. Gilles doesn't agree as he has known good harvests without August *canicules*. 'What about the plumping of the grapes?' shouts Gilles. 'No good having hot dry weather if there's no gentle rain to give the juice to the grape!' I wonder what sort of singing voice Gilles has. He could surely match those of Pavarotti and Domingo.

Geoffroy has also told me the story of the meringues, along with many other tales of his childhood at the château and life in Gageac. The mulberry trees lining the road from Saussignac down to the château, for instance, were planted by his grandmother for silk worm breeding, the produce of which was sent off to Lyon to make silk. When he was young, they had orchards of plum trees and walnuts as well, which provided some of the income for the château, along with vines and the local farms. He has also recounted how the *chai* his grandmother had constructed only cost as much as her first car, which she bought at the same time. 'Labour was cheap in those days.'

He's told me, too, some of the earlier history of the château and how the de la Verrie family came to own it. It started life as a watchtower and prison, a local stronghold of the English cause in Aquitaine in the twelfth century. Its soldiers surveyed the valley of the Dordogne for possible attacks. By the fourteenth century it was a formidable fortress and belonged to the lord of Duras, an influential and passionate champion of the English.

Aquitaine, or Guyenne as it became known, had become the possession of the English crown through the marriage in 1152 of

Eleanor of Aquitaine to the man who became Henry II of England. A vast swathe of land between the Pyrenees and the Loire was joined to the realm of England, yet because of his possessions in France, the English king was a vassal of the French king under the feudal system.

A conflict of interests and loyalties had been established which was to last three hundred years. The famous battles – Crecy, Poitiers, Agincourt – were punctuation marks in an intermittent process of small-scale warfare between the followers of the English and French Kings. The countryside was a shifting patchwork of 'English' or 'French' towns and châteaux. The struggle was finally resolved in favour of the French at the last battle of the Hundred Years War at Castillon, only twenty minutes down the valley from Gageac and its château.

I am fascinated that the château should have been English. Geoffroy laughs. '*Mais oui. Mais, comment dirai-je* . . . they were troubled times. And they all changed their vests frequently.' Sometimes it was English, sometimes French, sometimes Catholic, sometimes Protestant. The English influence was beginning to wane when Duguesclin, a renowned knight and French to the core, laid siege to it. It finally surrendered in 1377. The Duke of Anjou created a barony for Gageac and gave it to the Comte Archambaud of Perigord who was a servant of the French cause.

The other tower and middle sections of the château were built in the fifteenth century. By the seventeenth century all the trades were represented in and around Gageac, with families of master masons, carpenters, stone cutters, roofers and *forgerons*. Equally, in the commune were weavers, wool carders, a butcher and an inn at the bottom of the hill. Agriculture was the principal activity and barrel makers thrived in the area.

During the Revolution, the château was badly damaged. The beautiful stone staircase was destroyed. Baron Doussaut de la Primaudière, an ancestor of the de la Verrie family and official of

Napoleon I's army, had a property at Ste Foy la Grande, 15 kilometres away, and eventually bought the château de Gageac.

'They had a *Vente à la Bougie*,' Geoffroy had said. I asked what that was. He laughed. '*Oh la! C'était une histoire! Comment dirai-je?* The auction was held in Bergerac and a candle was lit. When the flame went out the auction would be over. The Baron was on his way to the auction when he was attacked by marauders who didn't want him to buy the château. But he succeeded in fighting them off and arrived just in time to buy it. '*Tant mieux,*' laughs Geoffroy.

He has also recounted the story of his grandmother, Genevieve, mother of Pierre, who loved Gageac.

She arrived in Gageac by horse drawn carriage, two cherry red geldings carrying her up the hill towards her new home after her marriage. Locals lined the route, and at the gates of the château, two young girls presented her with garlands of flowers. 'She was very cultivated, gay and original. My grandfather died shortly after the war from the effects of poison gas, so my father never really knew him. My grandmother never left Gageac. She loved it and wanted my father to love it as much as she did.'

Geoffroy's house with its exterior stone staircase leading to the first floor is apparently older than the château, as is the small house in Eida's field where bread was made for the village, probably supplying the soldiers who were billeted in the original château.

People, Protestant or Catholics, English or French, have lived out the drama of their lives here for centuries. I feel their presence; in the solitude and calmness of the church and the château, and especially in the vines. Time and progress, wars and famines, all ultimately governed by the nature of the land.

Tonight Geoffroy is coming to dinner with Odile, and Christianne Basque, another friend. They have never met and I'm not at all sure it will work. It soon becomes evident that it is absolutely the right thing to do. I watch them chattering at the table, laughing, all talking at high speed and taking obvious pleasure in each other's company. Odile and Christianne are as charmed by his zest for life and simplicity of manner as I am.

Geoffroy knows very well the *quartier* that Odile lived in for fifteen years in Paris, Christianne knows the area of Antibes he lives in, they all know Marseille very well and the evening passes surprisingly quickly. He recounts journeys he has made around the world, thanks to Air France, for whom he works in Antibes, along with trips made around France and Italy on his BMW motor bike. All are recounted with a wry sense of humour and a passion for people and places.

'*Comment dirai-je?*' he says, looking at me with his sparkling eyes as he tells Christianne and Odile about his job as part of Air France ground staff at Nice. His voice is clear and articulate and his laughter infectious. Trying to follow their conversation as they skip from one subject to another makes my head hurt and by the end of the evening it's awash with new words, and mental notes to look up phrases, words, and expressions.

'*Salut*. Can you help?' It's Beatrice from the hardware store in Gardonne. 'Julien needs some help with his English. Not much, but he's stuck and has to hand in his homework tomorrow. He's on his way up to you now.'

Julien, Roland and Beatrice's son, arrives ten minutes later, armed with textbook, exercise books and the homework in question.

How am I going to explain the English language in French? As it turns out, there's no need.

His English is very good and no sooner has he started than he's finished. He's in a hurry, he says, as he's going to Bordeaux to hit the nightclubs with his brother, who's a student there. His mother said he couldn't go if his homework wasn't done. He'll probably find his father in the same nightclub later on and they'll be in competition for the same girls, he tells me, laughing. Dad loves to dance, he continues.

You mean he drives all the way to Bordeaux for the nightclubs? I ask. It's at least an hour away. Aren't there any nightclubs nearer?

'Yep, but we've done them all to death. And anyway, it's good for a change of scene and a change of talent.'

The grapes have started to change colour. They are now much larger; swollen, yet compact. The colour change is gradual on the reds. First two or three on the outer edge of the bunch, then more and more. The whites have become less green and are slightly translucent.

We have stopped spraying as the grapes are no longer at risk from bugs and also, Gilles tells me, so that we don't poison potential customers. The grapes will not be clean again until a good month without spraying. I'm very happy to stop spraying, and even happier to stop trimming the vines. Having done it three times, I find it's no easier, no less stressful than the first attempt and just as dangerous as I originally thought. And I find I take it personally when the branches smack my face, shoulders and back.

Now we are all hoping for hot, dry weather, Gilles tells me. Yes, I agree, I am too. We need some rain, he says, but not much. Too much rain after hot weather will swell the grapes too quickly, break the skins and rot will set in, to say nothing of diluting the wine. Too much sun and the grapes won't have enough juice as the vines will have to use their water supply to keep cool, and it will restrict the ripeness.

Bruno, the *oenologue*, is back, to inspect the grapes and measure the sugar and acid content. We march up and down the vines. He passes a cursory glance at the reds, pronounces them clean and healthy and moves on to the whites, which will normally be picked first.

He explains that only when the grapes have arrived at ideal maturity can they be harvested. It is a very delicate balance; neither too early nor too late but only when the sugars have developed sufficiently and the acids have diminished enough. (Back to balance and the *dégustation* class with *tartrique* acids, *malique* acids and *citrique* acids in the sugar and *sulfurique* acid, *chlorydrique* acid and *phosphorique* acid elsewhere. 'A wine without acid is dull and insipid, a wine with too much is hard.')

'In the past, there was a feudal system which dictated the start of the vendange by a proclamation, *les bans des vendanges*', he says. 'Today, although there is still a *ban des vendanges*, scientific study of the grape and common sense dictate when to harvest.'

I can feel insecurity and fear galloping into my soul. 'How will I know?' I ask him.

'I'll tell you,' he laughs, 'but it won't be for a while yet, looking at your grapes.'

I look at them. They tell me nothing.

I am stressed. The vendange has begun in Gageac in earnest. Jean de la Verrie has already picked his whites and his merlot, along with Monsieur Cazin and Gilles.

'What's your *oenologue* thinking of?' shouts Gilles. 'They're all the same these technicians, white collar workers! He doesn't know your *terroir*, you should have picked your whites ages ago, and your merlot! So much for scientific study!'

The same was true of Monsieur Cazin. The *bans des vendanges* had been announced and with it had come the tractors, trailers, wooden half barrel, paniers and scissors. Once again, Gageac is filled with frenetic activity. Except for here.

Bruno visits twice a week to inspect the grapes. He takes a merlot grape, breaks the skin and looks at the pulp and the pip in the middle. 'No, it's not ripe yet.' He squeezes some juice onto a refracting machine and holds it up to the light to measure the sugar content. 'Mmm, *presque 13, quand même*. The bunches are healthy with no rot so we'll wait a few more days. The *météo* is good. See you Friday.'

Edge, Becky's brother, wrote earlier to say he would definitely be here for the vendange. He'd be no trouble and could help with the heavy work, being a rower and a fitness fanatic, he said. I drive to the station at Gardonne to pick him up.

Standing outside the station, he is incredibly tall and striking. Gardonne has probably never seen anything quite like him. Certainly Beatrice from the hardware shop has not. She is standing next to him, laughing and nodding as he practises his French on her. He is wearing an Australian leather hat with corks hanging around the brim, multi-coloured long shorts, no shoes and an orange tee-shirt with the sleeves ripped off. On the tee-shirt is written *Greenpeace Love Sex*. He has curly fair hair, framing an angular face, with a long nose and intense, clear and intelligent eyes. Although I know Becky as a longstanding friend of Chantal's and I have heard a lot about Edge, I have never met him. There can be no doubt who he is. The resemblance between him and his sister is very marked.

'Hi,' he says, and laughs. '*Ma Patronne?*' He looks from me to Beatrice and smiles, towering above us. Next to him is an enormous backpack and over his shoulders is slung a pair of Dr Marten boots, laces tied together as a strap.

Beatrice helps us manhandle the huge backpack into the back of the Diane. Edge climbs in to the seat next to me. He and his backpack fill the car, his knees almost touching the roof as we drive out of Gardonne and up the hill to Gageac.

I am vendanging tomorrow; the merlots. The white can wait until almost the end of the vendange, Bruno says. The grape is ugniblanc, usually used for cognac and only just tolerated under the *appellation contrôlée* rules. 'You could wait 'til almost Christmas for those to ripen,' he laughs. And indeed, compared to the semillon, muscadelle and sauvignon white grapes which will be used for the sweet wine and which are now pale yellow and transparent, these are huge, pendulous bunches of bright green, hardly a yellow grape in sight.

'Wow, Boss. Cool . . .' says Edge.

He is standing next to me as I inspect the merlot grapes. They do look very beautiful, plump and black. There are thousands of them, or so it seems as I look up the rows. My team will arrive tomorrow at 8 am to pick. Except for Michel, my team is comprised of novices. Roger is *hors de combat* again, as he harvested the row of vines at the bottom of his garden and drank the resulting wine in one gargantuan, bacchanalian feast. He has been dragged forcibly and with the aid of the gendarmes and the local doctor to hospital.

My team consists of Charles and Laura, over for a few weeks, Odile, Christianne Basque and her husband Richard, and three of their assorted offspring. Michel completes the team. And Edge. None of last year's team will be there as they belong to Richard Doughty and his vines and, sadly, none of the children are here, as they are all in foreign countries or working. James can't be here either.

'A lot of grapes . . .' Edge is saying. 'Cool. Are we picking them all in one hit, Boss?'

I have *paniers*, I have scissors and I have two *hottes* – large, elongated buckets with two leather straps that sit on the back of an unfortunate *vendangeur*. I also have the tractor, but no trailer. Gilles has lent me a flat wooden trailer, which is loaded with large black plastic dustbins.

I have cleaned and double cleaned the press. I have crouched inside it scrubbing the wood, I have stood outside it spraying disinfectant and water onto it. I have washed and steam-cleaned the crusher, the vats and, most particularly, the underground vat, where there were two decomposing mice on the floor, along with countless cobwebs full of spiders.

I am following Bruno's instructions to the letter. '*Chais* can't be too clean' he's told me. 'When you think it's clean, scrub it again.' All the pipes that attach to the pumps have had gallons of water and disinfectant sent through them.

'You'll never get the *malo* to start in this *chai*,' declares Gilles, 'it's far too clean. You need some microbes!'

I am cooking fifteen chicken legs, enough pommes dauphinoise to feed an army, courgette soup laced with Madame Cazin's milk, and apple tarts. Charles has found an old door next to the *marché couvert*. In a state of near collapse, this quaintly named amalgam of wood, breeze blocks and old doors at the bottom of the garden used to house the chickens and ducks when the previous owners were here. It is now home to garden tools and any household rubbish that can't be accommodated elsewhere.

Charles looks at the door. 'It'll make a table for six, so with your kitchen table we should be able to seat everyone,' he says 'I'm off to the hardware shop for some trestles. And you'd better show me how to use the tractor.' Charles is the designated tractor driver for tomorrow.

A large grey raincloud, troubling what was, until now, a clear, blue sky appears. The sun disappears.

'It'll be all right, you'll see,' says Laura supportively, watching me watching the sky. 'Relax.'

'Yeah, no sweat,' agrees Edge.

Morning. I open the shutters to Eida and the sun – it's not raining. But an hour later it's not looking nearly so bright. Why didn't I tell them all to turn up early, at 7 o'clock? I say to myself. We've already lost an hour.

'*Salut. Ça va?*' Roland, Beatrice's husband from the hardware shop, is here, with Michel. '*Je resterai jusqu'au midi,*' he says, smiling.

Christianne and Richard are already in wellingtons and their girls, Audrey and Carole, are leaning against the railings of the monument next to the church, laughing and joking. Peggy, Odile's daughter, is also there with Odile.

'Okay. Let's go!' declares Odile, who is also wearing wellingtons. As an afterthought she adds, 'What do we do?'

'Hey, wait for us!' Two pairs of neighbours have arrived – Pat and Gilbert Chaffurin and Wilf and Noreen Speakman, all of them part of the group outside the church after each service. Suddenly humbled by the good will and generosity of them all, I feel a lump rising in my throat.

Charles, Edge and the tractor have deposited black dustbins at various stages along the vines and we are picking, two to a row. 'No one leave their rows until every grape is picked on their side!' I find myself shouting. 'If you've finished your side before your partner, work up his side until you meet.' Edge has a *hotte* on his back, as has Richard Basque.

There is a cry of '*Panier!*' and Edge leans backwards and downwards to have a *panier* tipped into his *hotte*.

The bunches are clean, compact and beautiful and I can't quite believe we are picking at last. Charles is in charge of the tractor and he and Michel lift full black dustbin loads of grapes onto the flat trailer and deposit empty ones in rows that haven't yet been picked.

'Okay, Patricia! Now what?' shouts Charles. The trailer is heaving with full dustbins of grapes. We leave the vines and head back to the *chai* and the crusher.

The noise as I turn on the crusher is deafening. Charles and Edge tip in the heavy dustbin loads of grapes as it bursts into action while I climb to the top of the ladder to add sulphur to the must as it falls into the vat. Stalks spew out of the machine at one end while the grapes, skins, pulp and pips are hurled up a pipe and into a large vat.

The huge pipe shudders and shakes as the contents surge through it. We all three stare at it. Charles has hammered a series of large hooks into the wall, onto which we have tied the pipe to guide it from the crusher five metres up to the top of the vat. It looks very tenuous as the pipe vibrates and the heavy contents soar through it. 'Bit dodgy,' says Charles, 'but it's going to hold.' With a series of dull thuds load after load hits the bottom of the vat.

The sudden silence when the machine is switched off is impressive. I breathe a sigh of relief that the system worked without the pipe being ripped from its tethers, or the pump giving up. 'Well, that was all right then, wasn't it?' says Charles.

We pile the emptied dustbins back onto the trailer and race back to the workers. 'No bins!' they shout, 'Hurry up!' and we are picking and crushing and loading and emptying and it's lunchtime.

'It's 13.5!' I exclaim as I float the hydrometer in the juice we've just harvested to measure the sugar content. '*Oh là, là!*' responds Odile, then adds, '*Est-ce que c'est bon?*' 'Is that good?', looking from Michel to me. It is good, and everyone cheers.

'Hi Mum!' It's John ringing from Thailand. 'What's that noise? And how's the vendange going?' I feel a sudden surge of joy at the sound of his voice. 'Wish I was there, Mum,' he continues. 'Sounds as though you have a lot of people, though!'

All the *vendangeurs* are in my kitchen, having lunch and chattering at deafening pitch. The soup is hot and spirits are high. Edge is attempting a conversation with Michel and asking for more soup. 'Hot work, Boss. Need more energy food,' he says. Michel is rubbing garlic on his bread and explaining to him the health giving properties of it. Richard is talking animatedly to Laura, who listens politely and nods her head in appreciation in spite of not understanding any of what he is saying, and Odile is practising her English on Charles. Christianne is dishing out chicken legs and potatoes and wine glasses are filled, emptied and filled again. Pia, our Swedish neighbour from Saussignac, arrives at the door. 'Hi,' she says, 'I've come to pick.'

The workforce is definitely sluggish in the afternoon as the effects of the food and wine take their toll. Roland has decided to stay on and help, even though he ought to be erecting a conservatory. His energy is not diminished by the excesses of lunch and he picks grapes and lifts heavy dustbins of them with gusto, joking and chatting to the girls. Edge is also still going strong, marching up and down the rows shouting '*panier!*' gathering bucket load after bucket load in his *hotte*. He has stuffed newspaper between the straps of the *hotte* and his back to ease the pressure. Richard Basque is also suffering with the weight of the *hotte*. I can see him wince as each *panier* load is tipped in. By late afternoon my trusty team perks up again as rain clouds gather with only six rows remaining. By 6.30 pm the final row in the *parcelle* is picked.

Charles, Roland and Edge heave the last of the dustbins onto the trailer and at the *chai* deposit them in the crusher. Laura is washing *paniers* and scissors, and has neatly propped them in a row along the wall. Wheelbarrow-loads of stalks are deposited in a pile by the road and the end is in sight.

'I'll park the tractor round the back,' shouts Charles and drives out onto the road, turning sharp left past the church and back into the vines again.

'It's fantastic that we managed to pick so much', I say to Edge, feeling exhilarated and relieved. 'Thank you so much.'

'No sweat, Boss. It went well, didn't it?' he replies. He is sitting on the low wall in front of the house, a temporary respite before we head back into the *chai* to deal with the day's harvest. It is still hot and the evening sun sends its rays down onto the wall and Edge. He is sharpening the grape scissors in preparation for tomorrow's pick.

'Did you say it was church day tomorrow?' he asks nonchalantly. Yes, I say. 'Yeah, thought so.' And, after a moment or two he adds, 'Umm,' as he continues to sharpen the scissors. He smiles at me, his large eyes moving from me to the *monument du mort* in front of the church. I look at him, then at the monument. It takes me a moment to register what I am seeing.

The surrounding wrought iron fence is now a twisted heap on the ground. 'Umm, think Charles just tapped it. Only lightly . . .' he adds, 'with the corner of the trailer.'

Bruno has told me to do a *remontage*, that is pump over the wine as soon as the vendange is in the vat, and each morning and evening thereafter. I have pipes and pumps at the ready. I attach one long pipe to the bottom tap of the vat and to the pump, and climb the ladder with the other pipe over my shoulder. At the top I sink it into the vendange. Charles opens the tap and switches on the pump.

The initial force as grapes, skins, pips and juice gush through the pipes and back into the vat almost knocks me off the ladder. It quickly slows, however, as the pips and skins block the pump, which groans with the effort. I smell burning.

The pump gives up and the lights go out.

At midnight, the electricity is restored and the pump is cooled but still no headway has been made with the *remontage*. We decide to try and unblock the tap by pushing an untwisted wire coat hanger into it to move the mass of pips. How do other people do their *remontage*, for God's sake?

By 3 am, tired and demoralised, we calculate that in spite of stopping and starting the pump every five minutes to unblock the tap, we have probably circulated the wine for twenty minutes. Bruno had stipulated twenty-five.

'Hi, it's Sophie.' It is a message on the answerphone. 'Just to see how the vendange went. Loads of love.'

'You need a *fagot*,' says Roland the next morning. He disappears into the vines and returns with a bundle of sarments, the woody branches that have grown up the trellis of the vines. He wraps them tightly with string, then chops the branches at either end to make a neat bundle, a foot or so in length. 'Which vat are you using today?' he asks, and places the *fagot* in front of the bottom tap, securing it with a huge stone. '*Eh voilà*' and we have a filtering system.

This does not, however, help with the current problem of the blocked tap this morning. 'Try the middle tap,' he suggests.

I heave a pipe onto my shoulder and climb the ladder again.

It's slightly better, but time is running out and my vendangeurs are arriving.

Charles has brought the tractor round again and Edge is piling dustbins, *paniers* and *hottes* onto the trailer. It's Sunday and the wrought iron fence surrounding the monument is still in a heap on the ground. The priest has arrived and is gazing down at the damage.

My elation from yesterday's successful vendange has ebbed away and I now feel stressed and uncertain. I haven't managed one effective *remontage*. What will Bruno say? And what about all the pips that we've deposited on the floor of the *chai* each time we unblocked the tap? Don't I need them for the wine?

What will the priest think about us working on a Sunday? What will the Gageaçois who are arriving for Mass think about the demolition of their wrought iron fence?

Michel arrives. '*Ce n'est pas grave,*' he says. He will take away the wrought iron and fix it this evening. By tomorrow it will be just as it was. 'And once you add the yeast to the wine and it starts fermenting, the *remontage* will be a piece of cake,' he continues. I am not so sure of either. 'On y va?' he says and we're off again.

We are back with the first load for the crusher. Everywhere is pipes, pumps, noise and stalks. In a bucket, I mix the packet of yeast with some sugar and water. Bruno has told me to add it with the third *remontage*. In the time it takes to send the load of grapes through the crusher, the bucket is overflowing with fermenting yeast. No time to waste.

Charles and Edge pile empty dustbins onto the trailer and Charles heads back to the vines with them. Edge, armed with the coat hanger, opens the tap at the bottom of the merlot vat and switches on the pump so that I can add the yeast from above while the vendange circulates.

At the top of the ladder looking into the vat, everything has changed. Where there was a jumble of pips, pulp and pale juice, there are now skins, millions of them, forming a cap on top of the must. And the juice surges through the pipe easily, without help from the coat hanger. It has changed colour to a much darker pink and smells very different. It's fermenting. Does this mean I don't add the yeast, I wonder?

'I'd just add it,' says Edge. 'Give it a fighting chance.'

I throw the yeast in and twenty-five minutes of freely flowing juice later, the first effective pump-over is done.

Back to the vines and the grapes are coming in thick and fast. Dustbins are emptied, then refilled. A second pump-over and the juice is darker still. Lunch, more grapes and it's 6 o'clock. There are still ten rows left and it's starting to rain. The *vendangeurs* decide to carry on and finish them. The sun sets as we finish the last row by the light of the moon.

My team are utterly exhausted. Some can hardly walk, having

spent most of the day bending down to the low vines. Pia had tried doing press-ups as she reached the end of each row to ease the stiffness, another *vendangeur* had finished picking on her knees. Everyone is relieved that there will be no more picking for a week or so.

They fall into their cars and disappear up the road. No doubt home to hot baths, food and bed. Not so us.

As I enter the *chai* with the last load, the aroma is astounding – wonderful red fruit flavours. The vat holding the juice from the first pick is bubbling and the cap of skins has risen by at least a foot. Another pump-over and the colour is dark, dark red. Bruno has told me to be sure to spray the cap of skins well to prevent infections, but mostly to extract as much colour and tannins as possible as the colour resides in the skins. Only with skin contact does the wine change colour, he had said. And it has, along with my hands, which are now black.

We start the pump-over of today's pick, which works, thanks to Roland's *fagot*. We now have three large vats full of red grapes. The *chai* is a mass of pipes, pumps, buckets and dustbins full of water. Edge does the pump-over of the second vat with another set of pipes and a pump. After each pump-over we send gallons of water through the pipes and pumps to clean them before changing vats. The new floor now has small black stains on it from the skins that fall out of the pipes after the pump-over. They deposit their colour anywhere they drop, in spite of constant brushing, swilling and sweeping.

It is 2 am by the time the pipes are hanging up again, the floor is clean, the pumps are put away and we're finished.

'How much higher d'you think that cap of skins is going to rise?' asks Charles, looking at the first vat which is in full fermentation.

'It's the carbon dioxide that does it,' says Edge. 'I'm going to bed.'

The white owl breathes heavily as we pass the lime tree, exhausted and in need of sleep.

A faint aroma of cherries and raspberries pervades the house as I descend the stairs the next morning. Near the *chai* it's even stronger and in the *chai* it's overpowering. The cap of skins on the first vat has risen over the top and is cascading down the sides to the floor.

Two nightmarish hours later, we have pumped out some of the juice from the overflowing vat into a smaller one, gathered up the skins from the floor, washed down the outside of the vat and not even begun to do any pump-overs.

Bruno is here. '*Très bien, jolie couleur,*' he says. 'Yes, the answer is not to fill your 100 hectolitre vats to more than 70 hectos,' he says, when I recount the story. I hardly needed to tell him, as the colour on the floor and the half-full smaller vat told its own story. 'It's the yeast bodies eating the sugar and turning it to alcohol that creates the heat, and the gas that makes it rise.'

We are learning fast.

'If you want some rosé,' he continues, 'you must run off some of the juice from the other two vats now.' The grapes have been macerating overnight on their skins and are beginning to create some colour, a beautiful, pale pink hue. 'But be sure to separate the skins from the juice,' he warns

Separating the skins from the juice requires a black dustbin. And a large sieve to sit on top it. And a small pipe running from the tap of the vat to the sieve. And an empty vat. Add to this the usual two pipes and a pump to send the drained pink juice from the bin to an empty vat. We gather all the constituent parts and prepare to start. The pump-over of all three vats was due, but could wait the fifteen minutes or so it would take to drain off the rosé.

When we open the tap, beautiful pink juice runs through the sieve and into the dustbin, followed immediately by kilos of skins and pips which instantly fill the large sieve. We turn the tap off and get the buckets out. Charles and I scoop handfuls of pips and skins into the buckets.

After two hours of endless trips up and down the ladder carrying bucket-loads of pips and skins to throw back into the by now less than full vat, we have 40 hectolitres of juice.

'That's it!' I exclaim. We don't want any more!' I had thought we did, but not after all the effort and the time lost. We haven't even started the pump-overs of the three vats, to say nothing of the yeast preparations. And we have to clean up the mess we've just created. Why did I want any rosé anyway? It's more work, more stress, and one more vat – and yes, I'm running out of them.

The day passes in a confusion of pipes, pumps and *remontages*. The *chai* is awash with yeasts swelling in various buckets. Pipes criss-cross the floor and all three pumps are in use. Dustbins full of water ready for cleaning add to the confusion, and we trip over wires and pipes. There aren't enough hoses for cleaning, there is not enough water pressure anyway and we're running out of vats.

The first vat of red wine is now black with extraordinary aromas. It has a pump-over three times a day. Three times a day I take its temperature and its density, the degree at which the yeast is eating the sugar and turning it to alcohol. The quicker the temperature rises the better the extraction, ensuring a more active fermentation, says Bruno. Which will last for two weeks or so, perhaps longer. I contemplate two weeks of turning over three vats three times a day, plus the ancillary work with alarm.

The week passes in a blur of activity and chaos. I've performed a *débourbage* on the rosé, nothing to do with burgeoning buds, I discover, just more pipes and pumps. In fact it is a racking of sorts, a cold settling, removing the larger particles and solid matter that fall to the bottom of the vat. I seem to spend my life up the ladder. I put into the rosé a different type of yeast from that in the reds. And now it's bubbling away like the reds, with aromas of raspberries and strawberries filling the *chai*.

'Oh, que c'est bon!' exclaims Odile. 'Two bunches and my *panier* is full. *Panier!'*

We are picking the white ugniblanc and the bunches are huge.

'Giant grapes, Boss,' comments Edge. 'What did you give them to make them grow like that? Can I have some?' There are only five rows, but five very long rows, which disappear over the brow of the hill and down the other side. 'Tall vines too – dead easy.' It takes sixteen people only one hour to pick the lot.

We are back to the *chai* and the press. The grapes will be pressed directly, instead of going through the *égrappoir*. Edge carries heavy dustbins from the trailer on his shoulder and tips them into the *pressoir* effortlessly. Charles is impressed.

'I'm in training. Going to try the márathon this year,' says Edge with a nonchalant glance in my direction. 'The Boss gave me some of that stuff she gave to these grapes for increased muscle power and energy.'

The press fills up alarmingly quickly, with only half the dustbins emptied into it. Charles and I hurriedly position the tray under the press as juice begins to gush through the wooden slats. Pipes and pumps are fitted, the doors are firmly shut and we're pressing.

The barrel of the press groans as the plates at either side wind their way into the centre, and juice pours out into the tray to be pumped away to a vat. The quantity of juice is extraordinary. In the meantime, dustbins, scissors and *paniers* are washed. Pumpovers of the reds are done, and the pipes and pumps are cleaned while the press winds itself in and out, squeezing out more and more juice.

*　　*　　*

Bruno is here and we taste the reds. The juice is no longer sweet grape juice, but richly coloured and tannic wine. We will stop pumping over three times a day and switch to one, he says, but leave the juice macerating on its skins for a bit more extraction.

It has almost finished its fermentation and the cap of skins on the top is no longer a thick, solid mass but a mixture of juice with skins floating in it. He smiles. 'Don't you now think it was worth waiting for?'

I do. It is rich, tannic and black.

The first press load of white grapes is finished and the skins are tipped out onto the floor. There are tons of them. We shovel up wheelbarrow load after wheelbarrow load and run it to the roadside. My blisters return with a vengeance. The press is given a cursory wash and Edge refills it with grapes.

It is the early hours of the morning by the time most of the clean-up is done, along with *remontages* and temperature taking. We are suddenly all immensely tired. We have to wait for the press to finish. Watching it turn round and inwards, then outwards, the juice cascading into the tray, I press the pump button periodically to send the juice into the vat. Charles is sitting on a bucket watching the turning press, in a trance. Edge, leaning against the white wine vat, is wet. All of us have black stained hands and grape stains on our clothes. We are longing for the end of the day. There is still the press to empty, grape-skins to shovel onto wheelbarrows and the press to clean.

As Edge looks on, wine seeps out over the top of the vat he is leaning against and onto his head. It's happened again.

'Okay everyone! Only the purely rotten and desiccated, nothing green, nothing swollen and nothing orange! Each grape is precious!' I announce.

'What's purely rotten and desiccated?' my team exclaim.

We are picking noble rot. Bruno has carefully explained in his quiet, precise way how I should pick noble rot. I listened intently, taking in every word – which I am now repeating to my trusty team. They are as perplexed as I was last year.

'But it's everything you told us not to do with the other grapes!' they exclaim.

I point to some perfect noble rot.

'We can't pick that,' says Odile. 'It's disgusting!'

'You mean we drink wine made from this stuff? Cool . . .' says Edge. He inspects them as you might some primitive dead insect.

They do look pretty disgusting. I explain that this particular type of rot gives high concentrations of sugar and a perfect balance of acidity. They are not convinced. When I tell them to put their buckets underneath the vine, to leave the bunch where it is and to only remove the grapes with noble rot on them, leaving the rest to develop for a later date, they are incredulous.

'*Oh, la la, c'est compliqué!*' says Christianne.

'I daren't pick,' says Jacques. 'What if I'm picking the wrong thing?'

We gather in a huddle around the first vine as I explain.

Roger is back and part of the team. He can't bring himself to leave grapes on the vine, he says. '*C'est le gachi!*', A waste. 'Did your monologue tell you to do this?'

I am at a loss. Monologue?' I peer at him, worried that he's already celebrating my vendange before it's over.

'You were telling me about your monologue the other day,' he says.

Laura sighs and asks about the significance of the 'orange'.

'It's probably a sign of vinegar fly,' I say, 'so if in doubt, smell and taste. If it smells of vinegar, throw it away, if not, taste it.'

'You've got to be joking,' mutters Charles.

I am as unsure about this as the rest of my team. We set to work.

'Is this right?' asks Odile, proffering her *panier*.

We settle into a rhythm. 'It doesn't matter how slowly you pick, the aim is to pick only noble rot,' I shout, inspecting *paniers*.

'*Oh la*, here comes the controller,' says Roland, laughing as I peer in *panier* after *panier*, verifying that the right type of rot is in them.

Roger picks up his panier as I approach him, and holds it to his chest. In it are whole bunches of grapes, green, pink, rotten and orange. 'I can't do it,' he says, looking directly at me, but unfocussed and swaying dangerously. '*C'est le gâchis.*'

Bruno walks towards us through the vines. '*Voilà ton monologue,*' slurs Roger. '*Demande à lui.*'

Bruno approaches, unaware that he's changed métiers and is no longer an *oenologue*.

By the end of the afternoon, most people have black and grey sticky patches around their mouths and on their noses, having tested and tasted frequently. '*Mmm, c'est délicieux!*' exclaims Richard.

'Pretty noble,' agrees Edge.

Roger's *panier* now holds the least grapes, but each one is a perfect example of noble rot. His pride knows no bounds, and he laughs, displaying black, decaying teeth covered in furry, noble rot. Bruno has stayed to pick too and we have enough for a press.

It is pressed immediately, like the dry white. As we tip in dustbin after dustbin, a white powder rises from the press. Perfect noble rot, says Bruno. We close the press and start it rolling. Unlike the dry white, which immediately ran with gallons of juice, it yields nothing for the first ten minutes or so. Slowly, a grey liquid drips through the slats. As it drops into the tray, I add the enzymes as per Bruno's instructions. Before long, the grey liquid falling from the press changes to an unctuous, golden honey-like substance. We taste it and it's truly delicious.

We are pressing the first merlot vat tomorrow. Before making my journey past the heavy breathing white owl to my bed, I climb down into the underground cement *cuve* to wash it out in preparation. By now cold and damp, I wash the walls and ceiling of the dark pit, pump out the resulting water, climb out with relief, then attach the longest pipe in the *chai* to the bottom tap of the merlot vat and into the cement *cuve*. Bruno ties the pipe to the rim of the underground *cuve* with rope. 'You can let it run all night', he says, 'and your skins and pips will be dry in the morning, ready for pressing.' He opens the tap and red wine rushes through the pipe and into the *cuve*.

In bed, each time I close my eyes, I see noble rot grapes. Singly. In bunches. In the press and hanging on the vine, their colours ranging from pale purple to pink, to brown, to orange. When sleep overtakes me, in my dreams I see red wine rushing into a cavernous hole, lost forever. Or covering the *chai* floor, the pipe having slipped. And white wine mingled with rosé and red wine spewing over the top of its vats.

In the *chai* this morning, everything is under control. The underground *cuve* is now full of the dark red wine. Not a drop is on the floor of the *chai* and nothing is dripping from the tap. We open the door of the vat. We are met with a wall of tightly packed, dry red skins and pips, ready for pressing.

Standing in front of the open door of the vat with a large shovel and fork, I dig out skins and pips and shovel them into the large dustbins, while Charles and Edge heave them over to the press and tip them in. They are heavy. Skins drop on the floor. We place two, then three dustbins around the open door and I keep shovelling and digging, skins falling into the bins in front of me and over the floor around me. I need longer arms, I need stronger muscles, I need help.

Charles takes over for ten bins' worth. It becomes more and

more difficult to extract the skins and pips from the vat. Edge takes a turn, then me.

'Don't fill them so full!' shouts Charles as I shovel. 'We can't lift them!' The skins are now wetter and heavier as the level descends.

It's obvious that I will have to climb into the vat to dig them out. I get in. Although the lid of the vat is open as well as the door, the gases inside are powerful and overwhelming. I dig and throw, dig and throw into the bins, taking a gulp of air out of the open door every so often. The skins, now mingled with pips, are very wet. How can pips be so heavy? I have pips and skins in my hair and on my face, in my wellies and up my sleeves.

'How many bins are left?' shouts Charles.

It seems impossible to cram any more into the now over full press. Edge looks at it, climbs up to the open top, takes off his wellies and gets in, stamping up and down to make room for the ten or so bins still left in the vat. I can no longer speak. I'm too busy gathering pips and skins and shovelling them out of the door and into bins. My arm muscles are screaming with the effort and the blisters on my hand sting. I am driven by one desire. I must empty this vat.

'Arrive at 7.30 pm for dinner at 8.' I am giving a *fête de vendange* for my pickers. This is a *gerbeboade*, what last year's team mentioned when they threw the bunch of flowers on the last trailer load of grapes. It is traditional to have one at the end of the vendange. The last pick of the noble rot was last week and I've booked a restaurant in Monbazillac, *Le Moulin de Malfourat*, and invited all pickers, plus husbands and wives.

My exhilaration at having finished the vendange slips away with the news that James can't come over; he says he's not feeling up to the journey.

My team of *vendangeurs* are transformed. Gone are the wellies and gloves. Instead they are stunningly scrubbed, clean and chic. When I arrive most people are already there, drinking aperitifs and chatting to each other. '*Paniers!*' shouts Odile, and there is laughter everywhere and anecdotes and food and wine.

I think of how hard they all worked, how there wasn't a moment's hesitation each time I called them. It simply wouldn't have been possible to have a vendange without them. I look now at their kind faces and know and feel their generosity of spirit.

'*Qu'est-ce que tu fais ce soir?*' Geoffroy is back from Antibes for the weekend. His usual smile is gone and his demeanour tells me that whatever I might have been doing this evening, I'm not any longer.

'Let's go and have a pizza,' I say.

His relationship with his long-term companion is over. He's desperately upset, they've been together for fifteen years. We eat our pizza and I listen sympathetically as he tells me about the final split.

He knew it was coming, he said, for years. Tears well out of his eyes, and he wipes them away with a large paper napkin. I don't know what to say, so I say nothing. He's going off to Prague next week, on his own.

'Wow, a turbo-charged babe!' exclaims Edge.

Serge's niece has come for help with her English homework. Edge is sitting outside, repairing and oiling broken vendange scissors as she arrives on her bike. She is shy, very beautiful and sixteen; I see what he means.

'*Bonjour,*' he says, standing in front of her smiling. He is wearing shorts very similar to those that Michel wears; multi-coloured and knee length. And his tee-shirt with the sleeves ripped out and *Rainbow Warrior* and *Green Power* sprayed on the front in dayglo green. On the back he has sprayed in orange *Save the Whales*. He

wears large unlaced Dr Marten's boots and has a Mohican haircut.

She is noticeably intimidated, being five feet or so in height,
compared to Edge's six foot four.

'I'll help her with her homework if you like,' he says, framing
the kitchen door and towering over her. 'How d'you say "you're
the dog's bollocks" in French?'

We have repeated the digging out of red grape skins twice and all
our red wine vats are now skinless. We are experts at digging, dustbin filling and emptying, wheelbarrow racing and all related activities. Furthermore, we now have finely toned muscles, both legs
and arms, as well as very black hands. Edge, Charles and Laura
are leaving, Edge to return to Putney in London and serious training for the marathon, rowing and work.

In the *chai*, every vat is now full and each needs some attention. I take their temperature, I take their density. I put my ear
to them to hear the yeast chomping away at the sugar, turning
it to alcohol. I climb up ladders and look into them. I spend most
of my days and some of my nights in there.

Bruno has explained that we are awaiting the second fermentation on the reds, the malolactic, where good bacteria eats bad.
It softens the green malic acid by transforming it into rounder
lactic acid. Until this happens, he says, the wine is very fragile and
can turn to vinegar. This I know very well from James's experience with the 1990 harvest. I am listening very carefully. To every
single word.

'These are volatile acids, so you have no control over them,'
he says. 'You can't make it happen, but you can create the conditions where it might.'

'What conditions?' I ask, horrified again.

He laughs. '*N'aie pas peur.*'

I do most definitely have fear.

'If the wine is less than nineteen degrees in temperature, the malolactic will never start.'

The wine is currently at twelve, I know, as I take its temperature often.

'You need a *canne chauffante*,' a heating rod, he says.

'Where am I going to get one?'

'Try Monsieur Roy at Foyenne Motoculture.'

Monsieur Roy sits in his office and is certainly king of all he surveys. He is large; very large.

'I'd like to buy a *canne chauffante*,' I say.

'*Ah bon. A canne chauffante,*' he repeats. 'What type?'

Oh God. He smiles at me and waits. 'I don't know.'

He smiles. '*Ah bon.*'

I return with a large heating rod, attached to a box containing a thermostat and a long copper tube. I must somehow fix the heating rod in the wine so that it is centred at mid to lower height. The copper tube must be submerged too, almost touching the base but not the *canne*. And the thermostat must be set at 19.6.

I must also not let air into the wine, otherwise the volatile acids will rise and we all know what that will lead to. A thick electric wire is attached to the *canne chauffante*, another to the thermostat and yet another is needed to join them both together and plug the whole into the electricity.

An hour later and I have only got as far as opening the thermostat box. In it is what looks like the inside of a radio. Next to a series of tiny wires and copper blobs is a wheel, which I have turned to almost 20.

I have also unravelled the copper tube carefully and measured the distance from the level of the wine at the top of the vat to the base of the vat. I have marked a line with chalk at the appropriate spot on both the thick black wire of the *canne* and also the copper tube to be sure they will both sit at the correct distance from the base, at mid to lower height. And I just know that it's not going to be as simple as dropping the *canne chauffante* in. It isn't.

First I have to take off the three-phase plug attached to the wire at the end of *canne chauffante*. Then lift the floating lid that sits on the wine, climb the ladder with the *canne chauffante* – plus the thermostat box, the copper tube and a large wooden spoon. Then balance on one foot and wrap the wires of the thermostat box and *canne* around the top rung of the ladder to secure it for long enough to lift the lid. Then drop the *canne* in and thread the electric wire through the hole in the middle of the lid on the side which floats on the wine.

I descend to collect the screws, bits of plug, screwdriver and a rough scribbled note to indicate which wires go where. Re-attaching the plug, I carefully thread the copper tube through the same hole to submerge it, then lower the floating lid to sit on the wine and balance the thermostat box on top.

I descend again, this time to wipe my brow and stop my legs shaking. Finally, I unwrap the large black electric wires from the top rung of the ladder and rewrap them around the wooden spoon at the chalk level I marked on the wire before I started. Resting the wooden spoon over the hole on the outside of the lid at the correct height, I connect the plugs to each other, then descend the ladder again and turn on the power.

'Bravo' says Bruno the following day when he comes to take samples.

Although I no longer keep putting my ear to the vat to check it is still fermenting, I now do so to check that the *canne chauffante* is working. It hums quietly and clicks off every so often when it reaches 20 degrees.

'Mm, think it might have started,' says Bruno, tasting the juice.

I taste it. 'How can you tell?'

'*Ça pétille au bout de la langue*', it tingles slightly on the tip of your tongue.

It does and it has.

The house at the bottom of the road opposite Geoffroy's is being renovated. It belongs to the commune and was the *ancien mairie*, then a school. Gilles tells me that when he was young, he went to school there. '*Oui, l'institutrice était toujours soûle,*' the teacher was always drunk. *Milledieu!* 'She managed to take the register,' he says, 'but not much else. She had a bottle of alcohol in a paper bag at her side, which she swigged from constantly. We were dreadful to her, total anarchy from morning to night. Of course, we never learned anything and eventually they took her off to a drying out clinic and we were sent to Saussignac. That's why I couldn't write, even at the age of twelve. None of us could. *Oui, c'était dûr*, it was hard. What with that and the catechism lessons. We used to have our catechism lessons in the château and it was Madame la Comtesse who gave them.

'In those days it was very different,' he says, 'still *la noblesse* and all that stuff. Not now of course. I don't tip my hat to anyone.' I look at his hat, a baseball cap with the logo of a brand of cat food on it. Yes, it's a democracy now, he says. But it's just the same. Except it's *les fonctionnaires* who are telling us what to do now, *Milledieu!* What do they know about vines!'

And he's off again, breaking the sound barrier with his voice.

Bruno's wife is the beautiful Claudie. Brunette, small with large green eyes and pearly white teeth, she, too, laughs as she speaks, a tinkling, enchanting voice. I have met her only once. I am invited to dinner. Their house is in Monbazillac, not far from Bergerac.

'Welcome,' she smiles.

'*Salut*,' says Bruno.

The hot foie gras with fried apples and fresh truffles cooked by Claudie that night remains fresh in my mind even now, as does the sweet wine from Bruno's cellar that accompanied it.

'*Mmm, ça va,*' says Claudie, passing a positive verdict on the *cuisson* of the foie gras. 'It must just kiss the frying pan,' she says, 'in, over and out. Not a second more or it's tough and ruined.'

This is not tough, nor ruined but divine. They've decided to have some vines too, to make a sweet wine, they tell me over dinner. Just nine hectares to start. Bruno will continue to work as an *oenologue*, but they've found some vines with the right *terroir* and why not give it a go?

The foie gras is followed by a rabbit stew, cheese, then a wondrous chocolate mousse. I know it's late when I leave because my eyes have started to cross in an effort to keep them open and I know I've eaten too much because I can hardly move from the chair. Or maybe it was the four different wines we tasted, or maybe an amalgam of food, wine, French and friendship.

Chapter 7

MY TRIPS TO AND FROM THE LAB INCREASE RADICALLY WITH THE onset of malolactic in two of my *cuves*. Now that the vendange is over, Bruno's visits to his customers are less frequent. As a result, groups of them, mostly large intimidating men, come to the lab, gathering round the counter depositing their samples for analysis and discussing what I imagine are deeply technical matters.

The low hum of voices as I climb the stairs to the lab still makes my heart sink and my legs consider an about turn, but the vision of two hundred hectolitres of vinegar is never far from my mind, nor the distillery lorry that I see fairly often around Gageac. The driver now waves ominously to me as he passes by. I long for the *malo* to be finished, long for the racking which I must do after that and long for the wine to be stable. Who would have thought I'd long to do a racking?

The group of men part to let me through with my samples. '*Salut*, Patricia,' says Bruno and steps forward to kiss me. '*Est-ce que tu connais Luc? Luc de Conti?*' and I'm introduced to one of them.

'*Bonjour.* Bruno's talked about you. We're just discussing our *dégustation* meeting.'

I look at Bruno.

'I've got my agent coming over tomorrow afternoon to taste some wines,' says Luc. 'Come over afterwards and let's try them together.'

My third vat refuses to start its *malos* and Bruno has suggested that I wait for one of the others to finish, then add those *malos* gathered from the racking. Which is how I learn that the *malos*, which at this moment are central to my life in the *chai*, sit in the pink leas that fall to the bottom of the finished vats.

The pink leas look and smell delicious. Sitting at the bottom of the finished merlot vat that I have just racked is a thick, yoghurt like mass of magenta liquid. It falls from the vat and into the bucket I've placed under the open tap in great globules; globules of creamy, heavenly malo-filled leas. I run up and down the ladder leaning next to the non-malo vat, tipping in bucket after bucket of the stuff.

Three hours later I finish scrubbing out the vat, the inside of which was covered with the dark red scale of *tartrique* acid. I'm wet again, my hair is soaked, my feet are sitting in water inside my wellies and water is running down my arms, but the vat is clean at least.

I found Luc's house only via a particularly circuitous route, mixing up Ribagnac, where he lives, with Rouffignac. His samples are delicious. We taste classic whites, barrelled whites, reds and rosés and find grapefruit, plums, blackcurrant, chocolate, and *tabac*. We have a similar taste and interest and I no longer feel intimidated by him.

'You must join our group of *dégustateurs*,' he says. 'We're meeting next Friday morning and tasting the reds. Bring a bottle of yours.'

I'm horrified. I haven't got a bottle, I say.

'Just draw some out of your vat,' he says.

Love is in the air. Geoffroy has returned from Prague where he's had the most wonderful time, he tells me. '*Tu est libre ce soir?* Let's go to the *l'Imparfait* for dinner.'

L'Imparfait is a restaurant in the old town of Bergerac. It specialises in oysters and fish and is situated in a small alley, not far from the statue of Cyrano de Bergerac. It takes its name from the nose of the statue, the imperfect nose of Cyrano. Outside are table and chairs, with large yellow umbrellas shading them from the sun. Inside it is intimate with pictures and posters of the Imperfect, Cyrano. The atmosphere is always warm and welcoming, partly because it's more informal and relaxed than most of the other restaurants in Bergerac, but mostly because of its owners, Jean and Annie. Jean is large, welcoming and expansive. Annie is blonde, small and discreet.

'*Comment dirai-je?*' says Geoffroy as we sit down under one of the yellow umbrellas. It is an unusually warm autumn evening. 'Prague is indescribably beautiful. The whole holiday was divine.' He tells me he's met someone wonderful.

I express faint surprise that he's recovered so quickly from the pain and misery of two weeks ago. '*Quand même,*' I say, having learned this very phrase from Geoffroy, along with '*comment dirai-je*' which I love using and indeed, mostly have to.

'Och,' he laughs, full of life and happiness.

The fillets of red mullet on a bed of fresh salad with lavender flowers are delicate and succulent, and the roast monkfish with *pleurottes* mushrooms, juice of *cépes* and tiny, fresh vegetables melts in my mouth.

Each year all *viticulteurs* in the commune must file a declaration of how much they have harvested. It must be handed in to the *mairie* in November. The declaration form is large – very large, with endless boxes. I filled mine in as best I could and delivered it on the last official day. An official letter arrives, informing me that I haven't filled in my form properly and must present myself to the *Services des Douanes*. '*Ooh la, attention!*' be careful, warns Gilles, '*Milledieu*, they're the worst!

The *Service des Douanes*, or Customs, is housed in a modern building on the outskirts of Bergerac. I arrive at the stipulated hour, with Gilles's warning ringing in my ears. 'It's never right! There's always some figure that doesn't add up! *C'est tellement compliquée!* It's so complicated!'

I arrive to find a dozen or so *viticulteurs* sitting in the waiting room, each holding a file like the one in my hand. Have they all made mistakes on their declarations? An hour later and I'm ushered in to meet Madame Dufour. On either side of her are mountains of files. She picks one from the top. It's a file on me and there, along with a dozen or so other documents, is a copy of my declaration.

She can't read it, she says, as the numerous alterations I've made to the figures on the top copy have obliterated the revised figures on the carbons. It is an official document and must be filled in properly. It will verify that I have the correct amount of wine for my hectarage so must be filled in '*comme il faut*'. Each wine type must be noted accurately, she says, along with the hectarage it's gathered from.

I'm listening. I'm repentant. I'm drowning in a sea of new information on forms, hectarage, hectolitres and local wine tax requirements.

An hour later and she has rectified all the figures and stamped them with the official *Douanes* seal. 'Don't worry,' she says, smiling, 'everyone has problems with them.'

The phone rings and it's Chantal, with 'something to tell me'. Odile is over for dinner.

'What?' I say.

Odile, hearing my reply and watching my face, mimes 'She's pregnant!', laughing.

But I'm no longer laughing. Chantal *is* pregnant. The test is positive and she's been for her first antenatal check. She has to take finals soon and is revising, feeling nauseous, tired and un-student like.

My mind is racing. She's twenty-one, not married and in her last year at college. She can't possibly manage. How can I help, being here in France? An abortion is not something she can consider, she had said. She will have the baby.

'Mum, I'm fine,' she says. I can hear in her quiet, trembling voice that she's anything but fine. I want to catch the first plane over to her.

'Another three weeks and I'll be over to stay,' she says.

I need to apply for my label. More forms to fill in, and I'm not even sure what it is I'm applying for.

'*Ouie, c'est fait exprès pour nous embêter!*' it's deliberately made to annoy us, shouts Gilles when I ask him how to fill in the form later. 'Not only are they going to decide whether or not you can have it', he warns, 'you have to pay them for the privilege! And it's not just the INAO!' he continues, now in full vent, 'What about the CIVRB and the FVB?'

Completely baffled with the battery of acronyms streaming from his mouth, but unable to stop the torrent, I listen. 'You have to pay them all*! Milledieu! Les fonctionnaires! Les syndicats! Les techniciens!* What do they know about it? And we pay their wages to boot!'

I wait for the tirade to subside. My head is hurting and it's full of thoughts of Chantal and babies and I've no idea what Gilles is talking about. 'What is the INAO and the FVB?' I ask.

There is more invective as the house reverberates with the sound of his booming voice. I guess I have to fill in some more forms and pay some more money.

Gilles, when he's calmed down, tells me the FVB is the Federation of the Vins de Bergerac and that one of their employees will come in person to my *chai* and take samples of each of my wines.

He's going to take three samples of each and give me one of them. One of the others is going to the lab to be analysed and the third to the Commission of the Interregional Confederation of the Wines of Bergerac, the CIVRB, to be tasted. It has to pass a quality control test, he says.

I'm not sure why I need my sample. 'Because you might disagree with their findings!' shouts Gilles, building up a head of steam again.

I'm sure I'm not going to do anything of the sort and decide against asking him to explain the INAO on the grounds that neither of us can stand the strain.

Bruno explains it to me instead. The French *appellation contrôlée* system was first adopted in 1936 and is essentially a quality control system. Each *appellation* has its own variations according to its soil, grape variety and vinification methods. Ultimately, it should protect the growers, Gilles included, he explains. The INAO or *Institut National de l'Appellation d'Origine* sets out the laws that govern each system. They are in fact decrees formulated by the growers, the INAO being the supervisory body.

Yes, but what's the label and the *agrément* that Madame Dufour and Gilles talked about? Patiently, Bruno explains that just because my vines are in the Bergerac region doesn't mean they will automatically be classed a Bergerac *Appellation Contrôlée* wine. They must pass a test each year and conform to the regulations set out by the INAO, both analytical and *gustative*. An '*agrément*' is the official document giving permission to call the wine AOC Bergerac and the 'label' is the request to ask for it, which, in turn, leads to the visit from the FVB. My head is hurting again.

'*Bonjour* Patricia.' Luc has stood up and is advancing towards me. From the cacophony of sound a second ago, there is now only silence.

'*Salut* Patricia,' Bruno calls out.

Of the thirty or so people in the *dégustation* room, they are the only two I recognise. It's cold. It's 9 o'clock on a winter's morning and I'm terrified again and intimidated and I'm the only woman here, dammit. I have with me two bottles of wine drawn from my vat and I really would rather be somewhere else.

'This is Patricia Atkinson, everybody,' says Luc. 'Come on in.'

People mutter *Bonjour*, someone takes my bottles from me, uncorks them, covers them in silver foil to be placed with the others and we are tasting. We give a mark out of ten and make notes on the wines. Occasionally people comment on a particularly good one or a bad one during the tasting. '*Oh la, est-ce que ça mérite un?* Is it worth even giving it one?' someone says at sample six. I feel sure it must be mine.

We taste them all, then someone to Luc's left asks us for our marks in order to find a mean for each wine. Luc then returns to the first sample. 'We gave it seven. What did we think of it?' he asks. He unwraps the bottle. 'Christian, it's yours,' he says. 'What method of vinification did you use?'

I am stricken by horror and fear. Obviously, every winemaker here is going to have to comment on their wines, including me and I don't know what my method is and mine is number six and I've suddenly forgotten all my French and I'm fervently hoping the ground will open up and swallow me.

We pass wine number six and it isn't mine. My relief is immense. A great gush of liberation rushes over me. My wine comes in around the middle to lower section under the means test. Bruno helps out with the description of my vinification methods when I'm asked about them. I discover that I'm using 'traditional' ones.

After the tasting people get up and mill around with a glass of one or other of the better wines, drinking it rather than spitting it out.

'So you're making wines at Gageac?' Christian Roche is talking

to me. 'It's pretty good for a first attempt, and in a bad year too.'
He has vines at Monbazillac and has taken them over from his
father. I recognise him as one of the group I see at the lab. Well
built, about thirty-five years old with dark hair framing a kind face,
he talks with an easy, friendly manner. He was born in Colombier,
he tells me. I nod, feigning familiarity with the region in general
and Colombier in particular. 'It's not far from Monbazillac. Call
by if you're in the neighbourhood. Come and taste in the *chai*.'

Gilles is going to teach me how to prune — *la taille*. I've bought
the special secateurs, in fact two pairs. One, the usual gardening
secateurs, the other a huge pair of long-handled scissors. I have
also bought a small saw.

'Okay, now look at the wood in front of you,' He says.

I am looking at lots of wood; there is the wooden vine trunk,
the wooden *lat* attached to the bottom wire and an array of other
woods that have grown up over the season, replete with tendrils,
all intertwined among the various wires.

'You're going to choose one for next season and you're going
to cut the rest off, leaving two spurs as a security measure in case
you break the *lat* you've chosen while tying-in. But, be careful!
It's not necessarily the thickest wood that gives the best fruit. Not
all wood is fruiting wood and there are a number of other things
to take into account.'

This is next year's harvest I'm messing with, I tell myself in
alarm. I choose one. '*Non. Il ne faut pas que la vigne monte.*' It mustn't
be one that causes the vine trunk to increase in height. 'You must
keep the vine low,' he warns, 'otherwise its sap will have too much
work to do to get to the grapes.'

I choose another one.

'Yes, but are you sure it's a flowering wood?'

No, I'm not. I have no idea and I'm never going to get the hang
of this.

* * *

I have my *agréments*. An official CIVRB letter arrives and in it are three packets. Each contains an analysis form, a small green card and a piece of paper announcing that the red, the rosé and some of the white have passed both the analysis and *dégustation* tests. They can be labelled as *appellation* wines. A percentage of the white must be listed as *vins de table*, as the huge bunches of ugniblanc grapes that we picked for the dry white are no longer acceptable as a variety in the *appellation* and I have too many of them.

Michel is beside me again in the vines. As usual, he's arrived from nowhere. I've grasped the elements of pruning and gingerly move from vine to vine, relieved at having made critical wood choices and pruned one, anxious at the prospect of the next. Still, I'm quicker today than I was yesterday. Gilles says I've got the hang of it and before I know where I am I'll be galloping along the rows. Galloping?

I watch Michel gallop, cutting with a sure hand. He explains that once you have chosen a *lat* on the vine in front of you, while you are cutting it you should have already made a decision on the *lat* for the next vine. 'Think ahead, look ahead,' he says. I do and it works, after a fashion, although not a galloping one.

'*Ah, Madame, bonjour!*' It's Madame Cholet on her way to the cemetery. She's dressed in a muted orange roll-necked jumper under a Crimplene dress. Over it is a multi-coloured nylon overall and on top of that a cotton apron tied round the back and front. A large brown cardigan sets off the ensemble.

The cavernous pocket of her apron is weighted down, and from it she takes out a handful of small, almost linking circular hooks. She is decrocheting, she says. The steel links are attached at various intervals along the two wires on the vines that are lifted in spring.

Some people take them off before pruning, she says, others

141

after, but certainly always before wood pulling. She's almost finished, so will start on wood pulling tomorrow.

I can feel anxiety coming on. I've hardly started the pruning, let alone decrocheting, to say nothing of wood pulling.

'It's a hard job, pruning, isn't it?' she continues, 'Are your feet not cold in those *bottes*?' as she looks down at my feet.

We watch her wander off through the vines in the direction of the cemetery. I note again the pronounced swing of her hips when she walks, like a young girl.

'I've already unlinked for you' says Michel, 'it's much easier to prune that way.'

A few days later Madame Cholet is passing through the vines as I prune. She presents me with two sets of hand-knitted toe warmers.

'Put them in your *bottes* before you put them on. You'll see how warm they keep you,' she says. '*Mon fils*, I've always made them for him. He doesn't get cold feet pruning. It hasn't stopped him being ill, though,' she says, and recounts how Gilles suffers from crumbling bones as a result of heaving stones up onto his trailer. 'Driving his tractor over rocky ground doesn't help either,' she adds.

I look over towards Gilles's land and can't see any stones. 'That's because he's heaved them all up on his land,' she says 'and as a result he's full of lumps of steel!'

Grappling with this concept and finding it unfathomable, I decide not to enquire further.

'*Bonjour, Madame.*' It's Madame Queyrou whose husband has vines abutting Gilles's and mine. She's pulling dead wood off the wires, her husband working two or three rows ahead of her, pruning. I keep meeting people in the vines. Obviously, you're never alone in a vine.

Yes, yes, they always take off the links first, she tells me, much easier to prune that way. She's come out for an hour or two to pull wood, she says. Her son works at the chemical factory at Gardonne to supplement his income from the vines. I've no doubt seen him around; dark, handsome, always busy? He works in the factory all week, then tends the vines weekends. They belong to the cave co-operative so they don't vinify, like me. 'How is it going?' she asks kindly.

I receive regular letters from Edge, usually entitled *Missive from the Man on the Edge*. Sometimes it's an account of a drunken evening. Or a trip to Paris for an international rugby match. Or news of his latest girlfriend and how he feels about her. Often he includes poems he's written for magazines.

All his letters begin '*Chère Patronne, comment ça va?*'

Occasionally I get a phone call from him too, '*Patronne, c'est votre ouvrier*'.

He works as a draftsman, but is self-employed. He'll make sure he's here for the next vendange, he announces in one of his letters. And can he bring his girlfriend too please? 'She's a good worker, 'because she's a good rower and a good runner.' Among other things, he adds. And he'll be in training again, so make sure there's plenty to eat.

Serge's brother works daily on the renovation of the house at the bottom of the road. Every couple of weeks, a delegation arrives to inspect the work in progress. Today, the mayor of Gageac, Monsieur de Madaillan, steps out of his old blue Mercedes and greets a few of his municipal councillors who have gathered there some time before, along with an electricity board official and sundry others. After much discussion, they get back into their cars and disappear.

Monsieur de Madaillan stops in front of me. He gets out of the car, kisses my hand and announces that work is going at such a pace on the house, I should have neighbours in it before the summer.

Monsieur de Madaillan is tall and handsome with a large white moustache and an aristocratic manner. In fact he's much taller than the average Frenchman in these parts. His family have been landowners in Rouillac for centuries and he is a marquis. His château at Rouillac, large and sprawling, sits on the ridge dominating the valley a kilometre or so from Gageac.

He has vines, forty hectares of them. Michel has recounted how he worked for him on his property from the age of fourteen until he was made *cantonnier*. Monsieur le Maire has always been kind to him and his family, he says, and as a result, he goes over to help in the vines every now and then. But it's a hopeless task. The vines are in a parlous state, says Michel. They only managed to lift every other row of the wires last spring, and as a result couldn't weed or mow or plough. They sprayed those rows of vines where they had managed to lift the wires, but only just. So they were full of mildew, oidium and other maladies.

Why are they in such a state? I ask.

'*Oh, la, c'est une histoire!*' says Michel, touching his nose and raising his eyebrows with a faint smile on his face. 'Monsieur de Madaillan loves life,' he says. 'The de Madaillans have always been mayors here,' he continues. 'Protestants, which is why you don't see him at Mass.' And it's true, he is never at Mass.

Unlike the de la Verrie family, all of whom attend every service. I, too, have started going and I rather like it.

I'm beginning to know the different families. Monsieur and Madame Cazin and their two children, François and Rosalind, for example. In her late teens, Rosalind has the same wide and rather beautiful face as her father, with dark hair and ruby red lips. Always smiling, with large, white teeth, she is a perfect advertisement for the nourishing benefits of the rich, fresh milk of Madame Cazin's cows. François, her brother, a year or so younger, with a mass of curly, dark hair, large brown eyes and an extreme timidity of manner, works every hour of his day in the vines.

And Jean and Elizabeth de la Verrie's children, their two sons as handsome as their sisters are beautiful. Xavier works at the CIVRB and his elder brother, Hervé, Geoffroy's godson, works for Christie's in Paris. The girls are lissom, with porcelain skins and wide smiles.

'*Vous allez bien?*' How is it going? smiles Jean de la Verrie. He extends his hand, his enormous, hard, working hand. Mine are pretty much working hands now too. My fingers are fatter than they were a year ago, due to unscrewing large pipes on and off pumps or having them immersed in cold water in the *chai*, coupled with general outdoor work.

And, of course, Madame de la Verrie, la Comtesse, mother of Jean and Geoffroy, with her husband, Pierre. In their discreet way, they dominate the gathering with their grace and simplicity. '*Ah, bonjour Madame*. How nice to see you here,' she says with a smile. Her husband, Pierre, takes off his hat to greet people, a hat not dissimilar to the one his son, Jean, wears on his tractor. Victor, Jean de la Verrie's worker wears one too. Madame Queyrou is also there with her husband and granddaughter. '*Bonjour Patricia*,' and she gives me two kisses, one on either cheek.

I am not sure why I have started going to church. I love its serenity and I love living next to it. I love, too, the spirit it gives to the communal village. As there is no boulangerie, or shop or café, it is its very heart.

James is back for a visit. He is still unable to help as he feels constantly fatigued, and spends most of his time either in bed or lying on it. He found the journey here particularly tiring. And he really is not looking any better than when he was last here, which is some time ago. He now has a job, working for two weeks with a two-week break, which suits him well, given his low energy level, and he has a small flat in Stratford, east London, where his job is. His visit passes too quickly and he's gone, leaving early, as he needs to give himself time to recover from the journey before starting work again.

I feel depressed with each of his visits, which bring no improvement in his health or any alteration in our lives. I try to get through as much of the work outside as I can before his visits, so that I'm not working all the time during them. James, in his turn, tries to apply himself to the various forms and letters that accumulate in his absence. These, too, depress him. I have begun to open all letters addressed to him and deal with them. Most of them are either bills or forms or requests for information.

'*Oh la, la! Il ne faut pas tricher Patricia!*' Pepita is repentant. Pepita is mortified. Pepita is at my door. Deeply contrite, she explains that perhaps I'm a few eggs light in my basket. I buy two dozen a week from her and have never counted them. Madame Odile visited her just half an hour ago, she is saying. 'It's true, if you count them, Patricia, you'll find that perhaps there are less than two dozen. *Oh, la la!* But I sometimes make a mistake in the counting, and my weighing machine isn't what it was!'

Odile had bought ten chickens from her and paid for them by the kilo. Returning home, she'd weighed one of them before cooking to discover they didn't correspond to the kilos she'd paid for. Odile had put the chickens back into her car and returned instantly.

'But I've brought you some more eggs, laid this morning! *Oh, la la!* And you give Roger work!' Pepita tips the woolly hat back from over her eyes. '*Je m'excuse, Patricia!*' Her voice has the persistent, repetitious sound of a bird. '*Non, non, il ne faut pas tricher Patricia, non, il ne faut pas tricher . . .*'

The pruning is finished. Michel has come most days and worked for two or three hours, galloping up and down the vines in the fashion Gilles had told me I would. I haven't yet managed a trot, but am considerably improved. My right hand, wrist and arm ache. '*Oui, c'est la tendinites,*' smiles Michel. 'It'll go soon.'

And so to wood pulling.

'Stand to one side of the vine, not in front of it,' says Michel. 'Take the section of branch that has been cut from the vine with your right hand and pull it firmly down towards the ground but at the same time towards you with a sweeping movement. That way it will loosen itself and the accompanying branches from the wires and you'll be able to pull it out. *Et puis,* lay the wood in the middle of the row, not to the right or to the left. It'll be easier to *broyer le bois* later.'

I try. All the wires tense and the wood stays firmly where it is.

'You may have to cut a tendril, *une vrille,* or two to help it on its way,' he continues.

Where? Which tendrils? How will I know which one, I ask?

'Practice,' he says.

In the *chai,* the last vat has finished its malolactic and been racked and I'm about to put the reds into oak barrels. They are all second-hand barrels, new ones being '*hors de prix*'. Bruno has given instructions on how to clean and disinfect them and how to fill them. As usual in the *chai,* this requires water in abundance, pipes, pumps, time and patience.

Getting this far without losing the wine to the dreaded volatile acids is encouraging, to say the least. And so to work. I fill a barrel one-third with water from the hose and briskly roll it back and forth to move the water around and rinse it. Then I empty it and leave it to drain. Times fifty, which makes for wet clothes, toned muscles and a sense of achievement, to say nothing of a healthy appetite.

After draining, Bruno has emphasised how important the re-positioning of the barrel is. I must make sure they are absolutely aligned, so that they sit perfectly upright on the wooden rungs made for them. If not, the topping up, *'ouillage'* will always be imperfect.

I am making sure they are absolutely aligned and, working in rows of ten, feel it is deeply unfair that the eye can deceive so. Each time I think I have them all perfectly aligned, I see one that isn't and have to start again.

On to disinfecting the barrels. I need a wire coat hanger and a tiny sulphur tablet which looks like a yellow polo mint. The coat hanger is reshaped into a straight wire of approximately a foot in length, hooked at one end and with a flat circular handle at the other. I slot the polo onto the hooked end and light it with a match. An eerie purple flame spreads along the tablet and an acrid smell hits my nostrils with a force that brings tears to my eyes. I plunge the tablet and wire into the barrel, the flat circular handle on the outside, and place the bung over the hole. It burns for a minute or so.

Once it has burnt itself out, I must carefully remove it, making sure the husk of the tablet doesn't fall off and drop to the bottom of the barrel. As it is now mostly ash, this is very likely. In which case I will have to rinse the barrel again, which means moving them, and realigning them again.

There is no way I'm going to go through all that again. The care and precision with which I remove the wires and ash are second to none, and I succeed with all fifty tablets.

With the pipes and pumps in place, one pipe now leads from a full vat of wine, the other into a barrel. Attached to the end of the barrel pipe is a tap that I can turn on and off. Bruno suggested using the smaller pump, which is less powerful, and told me it takes approximately two minutes thirty-eight seconds to fill a barrel.

I have my watch with a second hand. I have opened the tap leading into the barrel. I have opened the vat tap. It only remains to switch on the pump and wait by the barrel until the allotted time passes, close the tap and move it and its pipe to the adjacent barrel.

Two minutes thirty-eight seconds pass. I turn the tap and wine shoots out of the barrel and hits the ceiling of the *chai* with unimaginable force. As it cascades down the wall and onto the barrels, it keeps on coming. I'm momentarily paralysed. The tap in the barrel won't close! I am momentarily paralysed. What to do first? The pump? The vat tap? Climb over the pipes and other barrels to get to them!

I scramble over the barrels, trip over the cumbersome pipe and eventually get to the pump to switch it off. The crescendo diminishes, the wine in the pipes hisses and the pump is silent. I leap back over the barrels and pipe to the overflowing barrel. As I lift the pipe and tap, yet more wine pours out before I can ram the tap into the adjacent barrel.

The pipes have stopped hissing and the only sound is of wine gently falling from the pipe and tap into the bottom of the second barrel. My earlier sense of achievement has vanished utterly at the sight of what looks like hectolitres of wine on the floor, over the barrels and on the wall. Wet with red wine and water and chastened by the experience, I set to with the hose to clean up.

It is midnight before all the barrels are full. I have refined my methods. The first barrel was forty-nine barrels' distance away from the pump. Now I have a longer pipe from vat to pump with the pump nearer so that I can see the barrel being filled. I also stand next to the pump and the on/off switch rather than in front of a non-operating tap in a full barrel. And I switch off the pump at two minutes thirty seconds. To hell with two minutes thirty-eight seconds. I top up the shortfall by hand with a watering can. Now I only have to clean the vats, the pipes, the pumps, the buckets and the funnels and I can go to bed.

My wood-pulling skills are improving radically and I now cut and pull with ease. More and more wood hits the ground in the middle of the row, in readiness for *broyer le bois*. Odile and Peggy come to give a hand. Peggy is a natural, Odile is willing. *'Oh la, c'est dûr!'* It's hard work, she says. 'How come you two can work so quickly?'

Madame Queyrou, my neighbour who has vines nearby, is not far away. She waves and shouts *'Bonjour Patricia! Tu galopes maintenant!'* *Oh la,* wait til Gilles hears this — *Milledieu*! I think.

Michel has invited me to supper. *'On va grignoter,'* come and have something to nibble. He lives in the commune of Gageac, but on the very edge, equidistant between Gageac and Gardonne. Monique, his wife, is younger than him and works at the gunpowder factory in Bergerac. They have one daughter, Marie Céline, much adored by them both, aged around sixteen. She is tall and willowy, with long hair and a pleasant smile. Gilles and his wife, Pamela, are also there.

The 'nibble' we have, cooked by Michel, is delicious. The soup is made from white beans and vegetables, with much garlic and pork skin, which is removed before serving. It tastes wonderful. Michel picked the vegetables earlier in the day. He shows me one of the white beans, peels it and gives it to me. It tastes fresh and young. The soup is followed by shoulder of wild boar, both cooked and shot by Michel. Marinated for two days in olive oil, garlic, onions and spices and cooked slowly in the oven, it is succulent and soft and full of flavour.

Gilles is president of the *chasse* and explains the reasoning behind the killing dates for wild boars, deer and so on. *'Milledieu*, you can't just shoot what you like when you like, you know!' he shouts.

'Gilles, calme. Calme, ta voix,' says Pamela. A pale redhead, with a gentle manner and sparkling eyes, she is the local postwoman and knows everyone. She laughs a lot and speaks softly. She and

Monique are very good friends and have known each other since childhood, as have Michel and Gilles.

Michel explains the culinary details of the wild boar dish. Gilles's way of marinating is different from Michel's, and Pamela's is different again. Quite a lot of the men do the cooking. Gilles cooks lunch in anticipation of Pamela's return on her post run each working day. He also cooks foie gras conserve each year, plus pheasant pâté and the like. Michel is known in the area to be a very good cook as well as a *chasseur* and *pêcheur*. He grows his own vegetables and has a row of vines to make his own wine; all these activities somehow fitted in along with his job as *cantonnier*, general Mister Fixit in the region and occasional galloper in my vines.

'In fact, it hit the front of the commune van,' he's saying, returning to the subject of the *chasse* and the wild boar. 'Just appeared from nowhere, like that!'

'*Milledieu*,' says Gilles softly. '*On ne sait jamais.*' You never know, do you?

'Did you cook it in that oven we bought?' Gilles asks him, a wry smile on his face. They all laugh and Gilles tells The Tale of the Oven. He had bought an oven for Pamela from a local supermarket specialising in white goods. He brought it home and set it up only to find that it didn't work. He rang the shop and spoke to the officious manageress who sold it to him. A faulty part, she said, they would send someone out. They didn't and he rang countless times, only to be rebuffed by the manageress.

'*Milledieu*, I waited for weeks! I'm a patient man, but enough was enough!' It was high summer and a hunting friend and he decided to take matters into their own hands. 'You remember, Monique? Bertrand, the woodcutter.'

'*Oh, la, oui, je me souviens,*' she said, laughing and looking at me meaningfully. 'They took the oven back, both of them', she continues, 'in their shorts, no tee-shirts. And Bertrand, with his beard and moustache and his huge pot belly!'

'*Oui*,' Gilles continues, with his wry smile, 'and I said 'where's the *gonzesse* who sold me this crap?' and banged it down on the counter.

'What's a *gonzesse*?' I ask.

'A *gonzesse*? You don't know what a *gonzesse* is?' Pamela, Michel and Monique are laughing.

'*Oh la!*' laughs Monique. 'A bit of stuff, maybe? The old broad?' she splutters. 'If you were being polite!' she adds.

'It's true she wasn't very nice,' continues Pamela. 'She played the holier than thou card every time Gilles rang up for the spare part. Didn't she, Gilles? All the same, can you imagine being confronted by those two, no tee-shirts, huge gut, booming voice, fags hanging out of their mouths?' Gales of laughter.

In the end, the *gonzesse* was only too happy to give them the spare part, their old oven and an entirely new one, which they later gave to Michel as a gift, to get rid of them, they were creating such a scene. 'Lowering the tone of the place, weren't you?' said Pamela.

'*Oui, c'était fait pour!*' It was deliberate! shouts Gilles, amid raucous laughter.

We eat the wild boar cooked in one of the ovens and drink a toast to the *gonzesse*.

Wood pulling is now almost finished. The days are beautiful, clear and crisp. I look out towards the church and beyond and see the cypress trees of the cemetery, dark green against the pale wall. The trees beyond are silhouetted along the ridge, bare of leaves but beautiful in form. Soon they'll burst into life again and so will the vines. I'm racing ahead with the wood pulling, but falling behind with mulching and, truth to tell, feeling a little anxious about what I'm sure will be another tractor lesson. I've forgotten what level of revs, what gear and what speed I must use in order to *broyer le bois*.

'*Il faut noter!*' pronounces Gilles, you must write it all down! Even seasoned experts like him forget from one year to the next, he says. Then 'second gear, full throttle, middle of the row, mower down gently, up at the end of the row' and I'm off again.

Chantal is here for two weeks. Looking pale and still feeling nauseous, she is not a picture of health. The baby is expected in August and, no, she doesn't want to know whether it's a boy or a girl but would prefer for it to be a surprise. In fact, she'd settle for simply having the nausea go away.

Sophie's been in touch from Boston, she says, and is sending her care parcels from America, consisting of books and bits and pieces to cheer her up. But she's gone off the chocolate Sophie sends as it only makes her feel sick. And she can't be bothered to read because she hasn't the energy.

In the *chai*, my barrels sit in neat rows and I now know what a *ouillage*, topping up, consists of, as I do it once a week. It's extraordinary how much wine the barrels drink for the first month or so. I top them up each Monday from a small vat of four hectolitres kept for the purpose.

Bruno has impressed upon me how essential it is to have a clean *chai* and I've taken his words to heart. Each time I top up the barrels, I take out the bungs and soak them in a water and sulphur dioxide solution. Each time I take wine from the tap of the vat, I wash down the tap. Then the floor. Then the jugs, the buckets and the pipette used for the final filling.

I now also understand how important it is to have the barrels aligned and sitting level. The wine must just touch the base of the round opening of the barrel, the opening that holds the bung. If the barrel isn't level, there's always an air pocket and, as I know from drilling by Bruno, 'where there's air, there's oxidation, where there's oxidation, there's the risk of volatility.' And I absolutely know what that means.

153

If the *ouillage* has been done correctly, the bung just touches the wine when replaced, excluding any air. A quick tap on each one with the wooden hammer and I can rest assured that all is safe for a week. But not before I have washed down each barrel to remove any drops of wine, as 'where there's air . . .' and even though it's on the outside, I don't want any there!

'You know, you don't need to scrub and wash down everything like that,' says Gilles, who's entered the *chai* just as I finish. '*Quand même, c'est propre,*' he adds. Yes, he says, he doesn't put his wine in barrels or bottles as he sells to a *négociant* directly. I've never visited Gilles's *chai* and, indeed, he never offers any advice on making the wine. 'No, my wine goes off to the *négociant* just after the malo,' he continues 'the sooner the better. The quicker he takes it, the sooner I'm paid.'

'*Bonjour Monsieur.*' It's Geoffroy, shaking hands with Gilles. He's just come back from Antibes. His smiling face tells me all is well. In fact, he is bubbling over with enthusiasm. '*Ah, je vois que les anges sont passés par ici,*' I see that the angels have called by, he says, looking at the barrels and the pipette in my hand and laughing. The evaporation in the barrels is the angels' share, he continues, the wine being so good they came to sample it in the night.

Geoffroy has made a decision. He has decided to retire early from Air France. They are offering a special retirement package and it would be silly not to take it. He would like to spend more time here in Gageac. His parents are getting older and the château and its environs are in need of attention. Perhaps he might travel more too, he says. His new liaison is going wonderfully and they will have more time to see places, do things and generally enjoy life. His face is the very picture of happiness. He's going to buy a new Peugeot to celebrate. '*Tu es libre ce soir? On peut aller dîner au Moulin peut-être?*'

*   *   *

Chantal is feeling much better, with the nausea almost gone. She looks beautiful again, but with an added glow. 'Don't be stupid, Mum, I look awful!' she insists. 'And it's back to finals. And you will be there for the birth, won't you?'

The white and the rosé wines are going to be bottled, along with last year's sweet wine. I must ring the bottler. Bruno says they are ready. The white is mediocre, we agree, but given that it's mostly ugniblanc grapes, it's not bad at all. The rosé is good and the sweet too. There's nothing to be gained by waiting.

The following morning a small, plump lady is at the door. '*Bonjour Madame*.' It is Madame Capponi, the wife of the bottler. She has a gentle voice, which matches her manner. She's come to find out how many bottles we will need. She sits down at the table and removes from her bag a large notepad. Then, with care, positions her glasses on her nose. In an instant, she calculates that 90 hectolitres is 12,000 bottles. What type of corks would I like, will she be supplying the bottles and do I think 'traditional' or 'simple' bottles are best? What about cartons and have I had the labels printed? Have I collected the capsules, the sleeves that cover the cork on the top of the bottle? And what about filtration?

Three weeks later and bottling day is imminent. Endless *échantillons* have been transported to the lab to check on protein, acidity, alcohol, copper, volatile acids, carbon dioxide, sulphur dioxide and most of the other substances that are found in wine. It is extraordinary that there are so many; the juice of the grapes is not such a simple affair.

A label has been printed, which is an achievement in itself. Rules and regulations apply to the label as much as to the making, growing and harvesting of the grapes. It must state the volume of the contents, the country of origin, the status of the wine, its *appellation*, the alcoholic strength in percentage of alcohol by volume, as well as the name of the producer. To say nothing of the *millésime*, the year it was picked and a lot number. Other, flowery descriptive terms are also admissible, the printer tells me. Don't I want to put details of how it was made? Wouldn't I like a print of my château on it? How about a back label as well?

Preferring the simplest label with the least amount on it, it does not have a château on it. Neither is there a back label. Apart from the legal requirements, it simply says *Clos d'Yvigne*, with the shield of Bergerac underneath it. The wine made by the previous owners was sold under the label, 'The Ancient Cuvée of Our Ancient Ancestors' which seemed both laboured and inappropriate.

The capsule trip was another lesson in the rigorous efficiency of French bureaucracy. The FVB, the Federation of the Vins de Bergerac, don't just send out their technicians to take samples when you apply for a label, and test them for their *agrément*, and take your money on behalf of the *syndicats* and the INAO. They also sell you capsules.

Capsules come in two categories: *congés* and *non congés*, tax and duty paid, and export. The export capsules are distinctive in one very important aspect; they don't cost nearly as much. Only *congés*, tax and duty paid capsules are sold at the FVB.

'How many would you like?' asks the lady in the office.

I don't know.

'Calculate how much wine you think you might export and how much you expect to sell here in France,' I'm told.

In confusion and embarrassment and having no idea whether I can sell any, let alone think about whether I can sell it in France or outside it, I decide on half and half.

'What's your *numéro de matriculation?*' she asks.

My God, what's that? I'd never heard of it. There are still great holes in my vocabulary.

It is, in fact, my car registration number . She makes a note of it then asks how long it will take me to get home. Mystified, I wonder whether she's decided I should not be let out on my own.

'I'll add ten minutes to give you time to stop at the *Régie* and *déposer* your *acquis*.'

Régie, acquis, déposer? She's probably right. I ought to get back home as soon as possible because my head hurts again and I have no idea what she's talking about.

Jean and Annie run a restaurant in La Ferriere at the bottom of the hill. They also have a café and sell cigarettes. And the café is also the *Régie*. A *régisseur* runs the *Régie*. My *acquis* is the document from the FVB. I must give it to Jean, who will then take some money from me – even though I have already paid some money at the FVB. He will then send the document to the INAO, even though the INAO offices are opposite the FVB office back in Bergerac where I've just been to get the document. And I must go straight home with the capsules.

The wine has been filtered. A man arrives to send the wine through his machine. Squares of what look like blotting paper are slotted into the middle of his machine which he changes twice before finishing. The wine, which was cloudy before, is now brilliant and sparkling. 'It only needed a gentle filtering,' he tells me. 'Much better for the wine.' Another visit to the lab with a final *échantillon* for analysis before bottling, and we are ready.

It is 5.30 am and through the vestiges of deep sleep I hear the sound of a large lorry and the clink of bottles. Then the sound of fork-lift trucks. I'm dreaming. I slept badly, knowing the bottling was scheduled for tomorrow at 8 o'clock in the morning. I open the shutters.

I'm not dreaming. A very large lorry is delivering 12,000 bottles at 5.30 in the morning. Fork-lift trucks are depositing them on the grass in front of the church. Eida is staring at them over the hedge. She looks up at me as the shutters open, as do the three workers. '*Bonjour!* We're just delivering the bottles, they shout. 'We'll be back later.'

It's half past seven and three more large lorries have arrived. Men shout as they position and reposition them in front of the house. Four people with wooden chocks in their hands are directing the lorries, shouting instructions to the drivers, 'No! Back a bit! Not too near the ditch! Start again!' The lorries are eventually positioned together, the chocks placed in front of their wheels, and they metamorphose into a bottling plant.

Two smaller vans arrive and park further up the road, blocking the church entrance and surrounding grass which is already littered with pallets of empty bottles. The roads out of Gageac are also blocked. Workers swarm around up the drive, into the *chai*, and onto fork-lift trucks, transporting bottles to the entrance of one of the lorries, attaching pipes to each other and depositing boxes of tools on the ground.

'*Bonjour.*' A large man is standing in front of me. 'Philippe Capponi.' He holds out his hand. 'Which are the vats with the wine in? We're starting with the white.' I show him, intimidated again by the hordes of unknown workers. 'It's not often we have a female winemaker,' he says and smirks, taking the cigarette out of his mouth and shouting instructions to one of the workers. 'My father will be here soon,' he continues. 'Ho, men! Let's get moving! There's work to be done here!' His voice is almost as powerful as Gilles's. 'Get that filter in place! Christophe! Over here with that!'

Pipes are attached to each other in the *chai*, the lid of the white wine vat is removed in an instant, and the Philippe in question already has a glass of wine in his hand, which he's holding up to the light. 'What filter number did you use?' he asks, then shouts 'Get the pipes set up to the filtering unit!' and the workers rush to do so. An irreversible decision has just been made to refilter the wine on its way into the bottle, as the previous light filtration is judged insufficient. Every available pipe is now linked up, leading from the vat, out of the *chai*, through a filtering machine and down the drive to the first lorry.

Mr Capponi senior arrives, pipe in mouth, cap on head. He reviews the scene, inspects the lorries then walks towards me and shakes my hand. '*Bonjour Madame. Monsieur Capponi*,' he says, with authority. People are everywhere; in the courtyard setting up the hose and filtering system, in the *chai*, in the lorries starting up the machines, all overseen by the hulking Philippe, who is fine-tuning a piece of equipment on one of the lorries while shouting instructions to his men. Mr Capponi joins him on the lorry, and with a raise of his hand, the conveyer belt on the lorry is set in motion.

Bottles rattle and shake as they move down the conveyer belt. They dance round, first empty, then full, then corked, capsuled, labelled, into a carton, off the belt and down to a worker who stacks them on a pallet, fifty cartons to each. Every so often, someone shouts 'Stop!' and the conveyer belt judders to a halt. A machine is slightly out of sync, requiring some adjustment. '*Merde!* Okay!' and the bottles move along the conveyer belt again. 'Change to export capsules!' 'No! Not those!' 'Change to rosé!'

The speed and efficiency with which they work as a team is impressive. Pallets, now full of cartons of wine, pile up side by side in front of the church. They have been wrapped in shrink wrap and glisten in the sun, shining white sentinels awaiting dispatch. For the moment, they will only be going as far as the atelier, the small work shop next to the house and already I can see there's not going to be enough room for them there. And they keep on coming.

Looking at a filled bottle with the Clos d'Yvigne label on it, I'm filled with delight and pride. It looks real and professional and the label doesn't look that bad after all.

'*C'est fini!*' shouts Philippe, and the conveyor belt stops, along with all the ancillary machines.

Silence and calm replaces the mayhem of the day. The workers group together, light cigarettes and chat softly. Then they jump down from the lorries and start unscrewing pipes, gathering up empty pallets, regrouping unused cartons and emptying capsule and cork machines. They swarm around, in and out of the *chai*, lifting pipes and washing them out, putting them and the used buckets back where they came from. The bottling plant, filtering machine and large pipes are hosed out.

'Where do you want the pallets?' asks a worker and the fork-lift trucks set to, lifting the full pallets of wine and heading towards the atelier.

'We can't fit them all in there!' shouts Philippe, glowering at me. I agree.

'*Oui, oui, c'est possible,*' decides the fork-lift truck driver. 'Move that one over to the left, then get rid of those tools,' and, with much moving and repositioning, the last pallet is eventually crammed into the atelier, a feat of unsurpassed determination and genius.

I sign a piece of paper handed to me by one of the workers, the lorries start up, the chocks are removed and they are gone.

I am shell-shocked, relieved and happy.

Ten minutes later and Bruno and Claudie are here. 'Just called to see how it went. *Ça va?*' says Bruno.

'*Oh, que c'est jolie!*' says Claudie, looking at one of the bottles sitting on the table.

There is a message on the answerphone. 'Mum! How's everything going? Love John.'

The red wine that only a short time ago was put into barrels now needs to come out again, albeit only temporarily. It needs aerating. The barrels mustn't stay empty for long, counsels Bruno, so I must work efficiently and fast.

Taking the wine out of the barrels requires a long, rigid stainless steel pipe with a series of holes at its base. It will sit in the barrel, attached to a pipe, of course, and the trusty pump. And another pipe. And an empty vat. The wine will be sucked through the holes at the base of the rigid pipe, through the pump and pipes and into the empty vat, leaving a residue in the bottom of the barrel. To remove the residue, I will need to turn the barrel upside down. Thereafter, the wine will be sent back into the barrels, but not before they are rinsed out and an analysis is done to see whether I need to add sulphur dioxide. And not before I have re-aligned all the barrels again. My heart sinks. Everything seems possible -except for re-aligning the barrels.

Starting at 6 am proves to be a good idea. Taking the wine out is not a problem. Rushing to the lab with an *échantillon* after ten barrels were pumped into the vat was a good idea too. Bruno promises to send the results back by fax after lunch. Finding a receptacle large enough to receive the residue from the barrels when I tip them upside down, yet low enough to fit under the rungs on which the barrels sit is more of a problem. Gauging where exactly the hole in the barrel is in relation to the receptacle once it is upside down with the opening no longer visible, is yet more challenging. I am deeply depressed as wine pours onto the floor rather than into its intended destination.

By the early hours of the morning the wine is safely back in the barrels. With great relief, I run a hot bath and soak my cold, tired and red-wine-stained body. As I lie there, I think about the events of the day and its achievements. At least the barrels aren't lying empty overnight. And I haven't lost too much wine in the draining, even though I seemed to. And taking the *échantillon* over to the lab after ten barrels were in the vat was timesaving too. *Échantillon!* I suddenly remember the *échantillon* I took to the lab this morning.

Leaping out of the bath and grabbing some clothes, I rush

downstairs, tears of frustration welling up in my eyes. How could I have forgotten to look at the results of the analysis? Already, in my mind, I am dressed, back in the *chai* and emptying some of each barrel back into a vat to add the sulphur.

The fax machine held the fax from Bruno, which has been there since midday. '*Pas de problème. Rien à ajouter. Bisous.*' Nothing to add, and big kisses.

Chapter 8

IT IS SPRING. IN THE SHADOW OF THE CHÂTEAU AND BEHIND JEAN de la Verrie's *chai* , the house at the end of the road is now finished and ready for habitation. It is, in fact, the only council house in Gageac. In England, we would die for a council house like this.

New neighbours have arrived. *'Bonjour Madame!'* Juliana is standing at her gate, listening carefully as I explain that I am her neighbour and have come to wish her *bienvenue* to Gageac. She is small, slim and in her late forties with long, dark hair and large, brown eyes. She replies and I understand nothing. Her husband appears at the gate and introduces himself as Yves d'Aureil de Palladin, speaking in exquisite French. He is older than her, tall, refined, and with impeccable manners.

Juliana is Italian, he explains. They have arrived from the Midi, where they have lived for the past four years. Before that, they lived in Italy, where he had a business selling marble.

'I sold the marble to your City of London,' he tells me. 'The marble for the new centre at Bishopsgate came from me.'

I wonder what a marble seller to the City of London is doing in the only council house in Gageac, or indeed, any house in Gageac. Is he still in marble, I ask him.

'Oh, a little, here and there,' he replies.

Juliana, meanwhile, is thanking me for the flowers I have brought for her, the few daffodils from the garden that have blossomed. Her voice is on a par with Gilles and Philippe in terms of volume. Each time I speak, she looks at me intently, with a pained expression, and I see in her a mirror image of myself only a year ago, trying to comprehend an unfathomable language.

'Yes, and are there any good restaurants around here?' asks Yves. 'And what are the people like in Gageac? We thought we might arrange a dance and open dinner outside once the weather improves.'

'*Voilà*. That would be magnificent,' says Juliana. 'Loud music, good food and dancing!' and she lifts up her arms, swings her hips and laughs, dancing round the garden. I try to imagine the people of Gageac dancing in the streets and can't.

Charles and Laura are back, for a week. They've come to look at houses to buy in the area. One in particular interests them, over in the Lot et Garonne. 'Hope you've got some work for me while I'm here?' says Charles.

Michel, who has arrived to mow the lawn around the church and has called in with a tin of home made pheasant pâté, smiles. '*Oh, oui, il y en aura*, yes there will be. I'll come and show you tomorrow.'

Replacing the rotten stakes and broken wires requires brute force. Walking through the vines testing each stake and noting which rows have broken wires is the first step. Testing the stake consists simply of shaking it. As a general rule, says Michel, if it moves in its hole, it needs to come out, so push it backwards and forwards to loosen it, take out the nails that are embedded to attach the wires to it, and pull it out.

Gathering the old stakes and delivering new ones is pretty easy with the tractor and trailer, as is untying broken wires and reeling them in. Actually putting the stakes, piquets, in the ground is another matter.

More than half a metre's worth of the piquet must be sunk underground to ensure a solid base for the vines. They must be knocked in manually with a large block hammer. After two hopelessly inadequate systems, Charles and I devise one that works. The first idea was to position a stepladder in the vines to give the height needed to supply the brute force to knock the piquet in. But the lack of level ground rendered this impossible. The second idea was a chair, equally inadequate. We decided eventually to drive the tractor up the rows with the mower attached behind, hydraulically raising it in front of each new stake so that Charles could stand on it, giving him sufficient height to swing the block and hammer the piquet home. I hold the piquet as Charles hammers.

Charles slept the sleep of the just that night, Laura tells me the next morning.

'Patriciiicia! Good morning!' booms Juliana's voice as I open my bedroom shutters. Her voice carries the 300 or so metres with ease. She is waving from one of the upstairs windows of her house. All her windows and shutters are open, with duvets and mats hanging out of them. 'What a beautiful day, *n'est-ce pas?*'

The dogs are standing to attention at the bottom of the garden, opposite her house, tails wagging. '*Quooqiloo, mes enfants!*' she shouts at them. 'Just a minute, I'm coming down to give you your breakfast!'

Eida is in her usual place, opposite my house awaiting breakfast, but with a startled look on her face. She looks over towards Juliana's house, then back at me. Things are changing around here.

Four pallets of wine are to be sent to England. James is trying to market the wine there. He's hoping to sell to restaurants as well as to our friends who have already ordered some. One pallet will be sent to Charles and Laura's for delivery to our friends, the rest will go to a small warehouse.

Bruno tells me that before the wine leaves, I must have permission from customs for it to travel, an *acquis*. And I know, by now, just where *acquis* are to be found. I drive down the hill to see Jean *le régisseur*.

'*Bonjour chère Madame,*' says Jean, a large smile on his face. 'What can I do for you this morning?'

I explain. '*Oh la!*' he laughs. 'An *acquis*? *Très difficile!*' and he reaches behind him for a large folder, full of forms. My gratitude knows no bounds as he fills the form in for me, saving me from the tortuous business of *congés*, volumes by alcohol, weight, code products, competent authorities and points of *sortie*. '*Le principal, c'est de la tamponner*', the main thing is to stamp it, he says, 'anywhere and everywhere, that's what they like!' and he proceeds to do so with a flourish, marking '*Recette Principale des Douanes de Perigueux*' and '*Correspondant Local des Douanes et Des Droits Indirects*' on all the six copies. '*Eh voilà!* One for luck!' as he stamps again on the top copy. '*Au revoir, jeune fille,*' and he laughs and waves as I leave, armed with the precious document.

The lorry, which will pick up the pallets of wine and transport them to England, is scheduled to arrive at 8 am tomorrow. I don't have a transpallet to get them to the front gate and, even if I did, it wouldn't work on the mud and gravel drive. I don't have a fork-lift truck either, so the only solution is to get up early and carry the cartons to the gate. Which is what I do.

It's 6 am and rain is falling. I have placed four empty pallets by the gate to stack the wine onto them. I race backwards and forwards to the atelier, carrying carton after carton to them as the rain beats down harder and the cartons soak up more and more water. I can see one of the builder's tarpaulins folded up next to the *chai* and decide to cover the half-filled pallet with it to give at least some protection to the boxes. I drag it to the road

Patricia in the vines

Patricia, Charles and Roland

Charles, Edge and Tim (left to right)

Edge

OVERLEAF
Above: The château at
Gageac et Rouillac
Below: Vista from the
vineyard over the
Dordogne river and
Bergerac

Left: Michel

Below: Roger

The *vendange* lunch at Clos d'Yvigne

Inside the *chai*

Gilles

Geoffroy

Juliana (left) and Odile

Clos d'Yvigne

and throw it over the pallet. Muddy water and sand pour out of it and onto the cartons. Miserable, soaked and racing against time, I continue until the 200 cartons are sitting on pallets by the road. I finish just as the lorry arrives. The driver looks at them, then at me and shakes his head, explaining that he has no transpallet or tailgate.

I find more empty pallets which the driver places on the lorry, then repeat the whole process. Together, we pile four pallets' worth onto the lorry. As it disappears down the road with its load, I walk back to the house, my arms still lifting unconsciously as a result of manhandling so many cartons, my spirits sunk in despair. Growing and tending the grapes and making the wine is as nothing compared to what I've just done. I feel tears welling up in my eyes.

I have been to the bank to see whether they can give us a loan. The new manageress, Madame Capin, is very sympathetic but not very encouraging. She will look at the figures and get back to me, she says. As far as I can see, there are no figures to look at. Our bank balance is almost at zero and there is no reason whatsoever why she should consider giving us a loan, there being no history of money coming into the account. Only money going out. We haven't sold any wine in the past as we didn't have any to sell and we don't have a business, therefore we have no business account.

She calls back the next morning to say that she can give us a loan with the wine that's in the *chai* as collateral. Can I bring her in my *déclaration de récolte*, along with various other official papers and she will calculate how much she can offer us? I feel a great load lift from my shoulders, along with an intense feeling of gratitude to her. Now I can pay some of the growing pile of bills sitting upstairs in the office.

*　　*　　*

'*Patricia, salut. C'est Luc.*' It's Luc de Conti on the phone. 'There's someone here you should talk to,'

Someone on the line is saying hello. The voice is English. It's a member of a Channel Four film crew. They're in the region and Luc has told them about me. Can they come over to see me?

'We'd like to film four English families who've come over here to do something in France,' says the director. 'Luc's told us about you and you're exactly what we're looking for. Would you be prepared to be filmed over a one-year period?'

I ring James who says I absolutely must. 'It'll be great for sales,' he says.

'*Patriiicia! Bonjour!*' Juliana's windows and shutters burst open. 'I've made some lasagne for you! I'll be over later! *Quooliqoo, mes enfants!*'

Sam and Luke have quickly formed a habit of being down at the bottom of the garden at the right moment and even Eida, after breakfast over the hedge here, has grudgingly presented herself to Juliana. The dogs' tails wag frantically in anticipation of Juliana's caresses and last night's left-overs.

Her small Yorkshire terrier barks maniacally. 'Chloe! Chloe! *Calme!*' she shouts at her as she continues barking. 'Oh, what a beautiful morning!' she cries, and duvets and mats are draped out of the bedroom window.

Geoffroy is going to install a small *appartement* in one of the towers of the château for his parents. It will be much more practical for them in winter, he says. He will do the work on it this spring.

Before that, he's going to have a small wall demolished next to his house. Would Serge be interested in doing it, do I think? Would I call him? He's been waiting for a month now for another builder and has given up on him. He would like it done sooner rather than later. '*Et, est-ce que tu es libre ce soir?*' Let's have a pizza.'

We have sold some of the sweet wine to Justerini & Brooks, a prestigious wine merchant in St James's founded in 1749. One of our friends in London, who knows them well and had once worked in the wine trade, has taken a sample to them and they like it. I am ecstatic. Justerini & Brooks! It's our first sale, and not just to a small restaurant and a few friends. The order is for two pallets and they'll be sending in an official notification soon, a fax informs me. And a lorry will arrive to collect the wine.

The lorry arrives in the afternoon on a beautiful day, dry and sunny. The driver accepts graciously the fact that he's going to have to work too, and I set about handing the cartons up to him. He's had some difficulty getting here on the small roads, he says.

'Attends! J'arrive!' shouts Juliana from her bedroom window. Soon she is handing cartons up to the driver as best she can, her cigarette clamped firmly in her mouth. Yves helps too and Michel, who's mowing the grass around the church.

From the bottom of the road I see Geoffroy waving. He too, runs up towards the lorry and the cartons. 'You must come and see the wall!' he says. Serge obviously had a spare morning to come so quickly. I hand him a carton, which he passes up to the driver.

Michel climbs onto the lorry to help the driver stack the cartons while we continue handing them up. They are on the lorry and shrink wrapped in no time.

'The wall . . .' says Geoffroy. The driver asks for the official papers as Geoffroy insists 'No, the wall! Don't let him go. Monsieur, you must come and see the wall.'

We walk down the road to the wall. It's still there and I wonder why we all had to come and see it now. 'No, not that one, the château!'

We gaze in horror at the thirteenth-century château wall, a

section of which has been utterly demolished by the lorry.

James has returned for a four-day visit. He's feeling slightly better, but still very tired. He has made the journey on his motorbike, to my dismay. Anyone would be exhausted trying to do such a trip in one day, I tell him. How crazy to do such a thing when you're ill. Geoffroy, however, is impressed, being a biker himself. *'Oh, la, tu es fou! Mais, c'est un joli moto.'* All the same, it's a nice bike.

The film crew is here. They consist of the director and his assistant, the cameraman and his assistant, the soundman and a rather beautiful girl. They are going to film the four chosen couples on the basis that have come to France to follow a dream. They want to film how reality compares with that dream. I consider whether I had a dream in the first place. I consider, too, the fact that I am no longer part of a couple in the real sense of the word, then decide not to follow that train of thought.

They're here to film my ordinary everyday life, they tell me. They have come over to look around and decide how to film it, they say. I'll soon forget they're there. And they won't interfere with what I'm doing. I'm just to carry on as normal.

The cameraman and his assistant are very tall so it's pretty difficult to forget they're here and the soundman has a huge furry brush on the end of a long pole that he waves about above my head. 'Can you put this on please?' he asks, and I'm wired up with a microphone. We all walk to the vines where I'm tying up the *lats*, a positive posse of people. I'm certainly not alone in the vines today.

'Could you take off your sunglasses, please, so we can see your eyes while we're filming you? And can you try not to squint?'

'No, Malcolm, sorry, we've got sound interference,' says the soundman as a distant tractor drives up and down rows mulching and 'Can we just do that again please? And can you look at me?' as I tie up a *lat*.

The *lats* this year are shorter. Bruno and I talk a lot about the state of the vines and the grapes; also the *terroir* and the wines in the *chai*. His support and encouragement, coupled with my, by now, very real interest and desire to know more spur me on, carton loading notwithstanding.

The most concentrated fruit is in the first three bunches nearest the foot of the vine, he tells me. The fewer the bunches, the greater the concentration in the wine. My *terroir* is *argilo calcaire*, lime based and has a dry, well-drained disposition which encourages deep vine roots. If water is near the surface, the roots will be shallow, if it is dry they will burrow down in search of water and minerals.

Vines produce better grapes on poor soil. As the soil can't produce enough food, its roots are forced further down. The old method in the area of ploughing fertilizer into the ground every year to produce a bigger and bigger crop is gradually disappearing, certainly amongst those *vignerons* who want to improve the quality of their produce. 'The way forward is quality, rather than quantity' he says. I did not buy fertilizer from the technician last year only because I couldn't afford it . I breathe a sigh of relief.

'There are three types of winemakers,' he is saying now. 'People like you who grow it, vinify it, bottle it and sell it. People like Gilles who grow it and vinify it to a basic level, then sell it on to a *négociant* who blends it with other wines and sells it under his own label. And people who grow it as a crop, then gather it in and take the grapes directly to the co-operative, who, in turn, make it and sell it for them. So,' he continues, 'as you're doing the whole process yourself, you might as well do it really well as not, don't you think?'

I do.

It's the early hours of the morning and I'm racking the barrels of red wine in the *chai* again. With only the red wine vat to clean now, I am armed with a hosepipe and water and a scrubbing brush. My arms are aching after so much barrel rinsing and re-aligning. I've finished filling the barrels and topping them up and I'm just too tired to do anything other than rinse out the vat cursorily, leaving the scrubbing and a more thorough cleaning for the morning. I pass by the heavy-breathing white owl and fall into bed.

Next morning, scrubbing off the tartaric acid encrusted on the sides of the vat I'm standing in, I curse and wish I hadn't left the cleaning until this morning. It's now much more difficult.

The film crew arrives and the cameraman gets very excited. 'Can you hold that?' he says. 'We'll just put some lights up,' His assistant rushes off and returns with huge lamps and lights, along with yards of electric wires.

The cameraman, perched precariously on top of an adjacent vat, is looking for good angles as I scrub and wash. Electric wires drop onto the wet floor and water from my hose sprays out of the vat door and towards the lights. Every so often, the beautiful girl shouts 'take two', claps a clapboard and they're filming. Getting out of the vat, I stand back and aim the hose at the door of the vat, the taps and the outer sides to finish the job. A scream as the beautiful girl appears from behind the vat, drenched and unhappy, having hidden there to be out of view of the camera. At the same moment, the lights go out as water from the hose shorts their lighting system.

*   *   *

I have to think about pulling up the ugniblanc white grapes. Bruno says they will never make anything other than a very mediocre wine. Ugniblanc is an Italian grape, used to make cognac. It never reaches a very high sugar content and always retains acid, usually in an unbalanced way, more acid than sugar.

'Your problem is that replanting is expensive and it will take four to five years before you have a harvest from your young vines,' he says.

'What do you want to pull them up for?' asks Odile incredulously. 'We love picking them. What a waste!'

A sense of impotence, frustration and depression overtakes me. I can't pull them up because I can't afford to replace them. Even if I could, it would be four years before I would have any harvest to speak of, and even then, only a small one. We would probably be bankrupt by then. It's a vicious circle.

Even more worrying, our only sale for the moment is to Justerini & Brooks. James has tried various outlets in London and the south east, to no avail. It's not easy to sell wine. In fact it's proving impossible. Apart from two wine bars near Tonbridge and our friends, it's been a complete failure.

'*Patriiicia! Bonjour!* I'm coming to help you *attaché* today!' Juliana's voice booms from her bedroom window to mine. Half an hour later, she is in the vines with me.

Michel has arrived too, as usual, from nowhere.

'Bonjour Madame' says Madame Cholet. She is on her way to the cemetery. Gilles's lumps of steel are giving him pain, she says, shaking her head. 'The unfairness of life. Those who have straws in their hands don't have this problem', and she makes her way to the cemetery.

'What straws? *Qu'est ce-qu'elle raconte?*' What's she on about, asks Juliana.

While we work along the rows tying up *lats*, Juliana describes her wonderful house in the Midi, large, with a swimming pool and acres of garden. The esprit of the people of the Midi is very different from here, she continues. There, you would have an open dinner any old time, and dancing in the street. The wine would flow, laughter would be heard everywhere; very different from here.

Why did she leave, I ask.

'*Oh, la! C'est une histoire!*' she exclaims. 'How much time have you got?'

They lost all their money on silly deals and the drop in marble prices. The bank and the bankruptcy court took everything; most of her jewels, even their beautiful Mercedes and now look what they've got, a tiny Peugeot! Then she laughs, her raucous, infectious laugh. '*Mais alors, ça fait marcher les commerçants!*' It keeps the shopkeepers in business. 'I'll just have to buy more!'

Most of my evenings are spent trying to comprehend the books Bruno has given me; *Le Vin de l'analyse à l'élaboration*, on understanding exactly what the analysis results mean, or *Fermentations, The Compounds in Wines and Inorganic and Organic Precipitates*. Or, more interestingly, in one of the books, an illustration of Bruno's *réfractomètre*, the small piece of equipment he carries to measure the sugar content of the grapes at the moment of the vendange.

It consists of two prisms. A drop of juice from the grape is held between the two and the light passing through it bends at a different angle, according to its sugar content. Through the eyepiece, a scale can be read giving the percentage of sugar in the juice, which saves picking two or three bunches of grapes, squeezing them into a receptacle and measuring the sugar content by the more traditional means of the hydrometer.

*   *   *

Gilles's neck and left arm are giving him great pain.

'*Oui, ce sont les chirurgiens, les médecins!*' he rails. He has crumbling vertebrae at the top of his neck and the base of his spine, due to overwork and heaving great stones around for most of his life. 'They're useless, these surgeons!' he continues, decibels rising. He'll have to stop carrying the stones around.

I suggest it might be a condition that exists independently of the stones and maybe he should go and see the doctor again.

'What do they know?' he explodes. ' I know what my body's doing! I've had two operations already! I've got two steel plates in my spine to show for it!'

I'm horrified.

'One of them's slipped, that's what it is!' He sees the books that Bruno has given me lying on the table and disgust crosses his face. 'And those books won't teach you anything either!'

A week later and he's back in the kitchen. '*Oui, c'est exactement ce que j'ai dit,*' it's exactly as I said, he pronounces. He must go into hospital for an operation to readjust the steel plate at the top of his vertebrae because the specialist thinks it has slipped. 'Oh, I'm used to it,' he continues. 'I'll be out in a day or so.'

I'm alarmed that he should be out so quickly. Isn't it a serious operation, I ask. Won't he be out of action for some time? And what about the vines? Can I help?

'Ha!' he laughs. 'You're very kind but my mother's already almost finished tying in the *lats* and she's the one who cuts off the shoots at ground level anyway. You just get on with your vines! And it's almost time you started your shoots and you've still got *attaché* to do!'

He's due to be admitted to Bordeaux hospital the following week. He'll be out within a week or so, but will have to wear a collar for six weeks. Pamela will look after things on the vine-yard.

I find the hospital, the same one that James spent time in. Called the Tripod, it's enormous. I find Tripod II and Gilles in a large building half a kilometre away from where I've parked the car. He is sharing a room with another patient. Lying on the bed with a large collar around his neck, he has to move his entire body to turn even slightly.

'*Salut*,' he says, smiling, genuinely pleased to see me.

Pamela is there too and laughs. 'What next!' she says softly.

The buds are appearing again, and the tender leaves are opening out. The soft clusters of pale green leaves with a rose tipped hue look so delicate and small that one can hardly credit what they will grow into. I think of the chaos of last year and the prospect of cutting off the shoots, lifting the wires, weeding, spraying and the horrendous trimming with a shudder.

The possible maladies the vines might catch are suddenly uppermost in my mind too. The book I have on virus and virus-like diseases is sitting with the other books Bruno has loaned me. A quick glance at the most important virus diseases is enough to frighten the most stout-hearted. Grapevine vein necrosis, leaf roll, black rot, red rot and *flavéscence dorée* and I decide it's time to shut the book. And yet, looking at the clusters of leaves, I feel a sense of excitement. In spite of the risk of virus, frost and hail. In spite of the forthcoming chaos, which I can see just around the corner.

Sophie is here. She's back from Boston for two weeks and is spending her last few days in Europe here. 'Poor man' she says when I relate the story of Gilles's neck. 'You ought to be careful on that tractor, you know. You might end up with the same thing.'

*　　*　　*

The red wine in the barrels is progressing. I can taste and see its evolution. I taste it regularly. Holding the glass up to the light and then looking down into it I can see its deep colour, black and rich red. Putting my nose into it, I can detect the aroma of cherries. Then, swirling the glass and aerating the wine, not just cherries, but blackberries and vanilla with a hint of spice. I look for bad notes as well as good and concentrate on acidity, bitterness, astringency and any vegetal note.

'*Salut Patricia.*' It's Luc, who's called in on the off-chance to taste. We taste and I feel proud and pleased when he gives his approbation.

Gilles is back. He walks up the road from the château towards me. He can't reach over to give the usual two kisses as a greeting. '*C'est emmerdant!*' he shouts in his usual way, and I know he's on the mend. He's got to walk on the road and not through the vines, he tells me, as the ground is too uneven.

And this time, the surgeon has told him not to get back on his tractor too soon. ' It's easy for him to say! Last time they said eight weeks, but you don't want to believe everything they say. I was back on it again within the week,' he says, smiling.

I'm appalled.

'It's okay, I'm not going to do that this time. All the same, it's *emmerdant!*'

I consider asking about the straw in some people's hands that his mother had talked about, then decide against it.

The film crew have gone back to England. They plan to come out every three months and do a couple of days' filming with each family, the director, Malcolm, tells me. I now know them slightly better. Graham is the cameraman, Stuart the soundman, and the girls, Bernice, Benetta and Jackie are cameraman assistant, director's assistant and general fix-it person. On their last day of filming, they put down their cameras and files and start helping, which they do with vigour for the rest of the day. Cutting off the shoots at the base of the vines is not for the fainthearted – or indeed the faint backed.

Another tasting, this time of dry whites, at David Fourtout's. His father's vineyard is situated on the ridge above Monbazillac, ten minutes' drive from me.

There are more people than usual here today. Christian Roche comes over to say hello, as does Luc. *'Ah, la viticultrice!'*

We taste the wines. I haven't brought along a sample of mine because it's ugniblanc. We discuss the merits or otherwise of one of the wines. 'Is there some residual sugar in this?' I ask shyly, then wish I hadn't as there's silence as everyone looks at me. The sweetness is evident on the tip of the tongue, in spite of the over-all dryness.

'There's some late harvest semillon in it maybe?' asks some-one else and an animated discussion begins. They include me in the discussion on the merits or otherwise of a percentage of late harvest semillon in dry whites and I feel, if not one of them, certainly not a total outsider.

We taste twenty or so wines. Some have gooseberry flavours, mingled with lemon, others rich fig and melon; good acidity in the best of them, less structure and complexity in the others, but all well made wines and each of them discussed and debated, analysed and marked.

'Yes, it's a good idea to pull up the ugniblanc,' says Luc after the tasting. Can't you buy any sauvignon vines around you?

I am introduced to Thierry, a great friend of David Fourtout's. They were at school together and tease each other mercilessly, even in the tastings. *'Oui, c'est le vin de David, j'en suis sûr!'* I'm sure it's David's wine. What does he put in it? You'd think he'd have learnt something by now, *n'est-ce pas* David?' and he laughs. 'You must come over and visit my *chai*, Patricia. Knocks spots off David's!'

Bruno and Claudie have rented their nine hectares of vines. They have them for an eight-year term, with the possibility of buying them at the end of that term if they wish. We are looking out from the terrace of their house as Bruno points at the middle distance. 'There they are.' They are predominantly semillon, with muscadelle and a small percentage of sauvignon. Under the appellation of Monbazillac, they will make only sweet wine.

'Have you finished cutting off the shoots?' asks Claudie. 'My leg muscles and my back!' she groans.

'It's great terroir' says Bruno as we continue to gaze out over the vines. 'It's hard work!' says Claudie.

'*Oui, tu sais, moi aussi, je suis comtesse!*' says Juliana. We're finishing the last two rows of *épamprage*, dressed in wellies, leggings and old jumpers. She is wearing a large diamond ring on one of her fingers and running a hand, the hand with the ring, across her brow. '*Oh là là, qu'il fait chaud!* In Italy, you're always too hot if you have a ring to show off!' She delicately runs her hand over her brow again, flashing her ring. She is wearing, in addition, some gold and pearl earrings and a large, gold necklace. 'One never knows!' she laughs. 'You told me you're never alone in the vines! *Eh voilà!* Here comes Gilles. Gilles!' she shouts, almost bursting one of my eardrums.

'*Salut!*' he shouts back as he advances towards us. He's still wearing his collar.

'I thought you weren't supposed to walk through the vines,' I say.

'*Milledieu!* That was ages ago. I'll be on my tractor tomorrow!' he shouts.

Everyone is shouting, me included, though I'm not quite

sure why as we're all standing together in the vines.

'*Oh, la*, it's warm' announces Juliana, laughing and running her hand over her brow, diamond sparkling.

Gilles is feeling much better and, although the six weeks is not up, he's going to do some tractor work tomorrow. It'll only be light work; no more stone gathering, he says.

Odile and Christianne are here for supper. 'It's obviously infectious', says Odile. 'What am I going to do?'

Peggy, Odile's daughter, is pregnant. She had a phone call from her earlier in the day, an almost identical phone call to the one I had from Chantal some months ago. Worry and stress show on Odile's face.

'At least Chantal's only in England; it's a long way to come back from la Réunion for a couple of week's love and attention from your *maman,*' she continues. Peggy is finishing her teacher training in French Reunion, off the coast of South Africa – or she was. Now she's feeling too sick, tired and pregnant to go to work. Odile is ready to pack her bags and take the first flight out, but Peggy insists she's being perfectly well looked after by her boyfriend.

Christianne and Richard's daughter, Virginie, is also pregnant, and has decided to marry her boyfriend.

'*Oh, la, les filles!* It isn't easy being a mother,' Odile says.

Christianne agrees. 'Let's have another *blanc cassis*. Here's to the babies.'

Spring is here. The wisteria in the courtyard is again a dazzling, pale lilac against the dark pink canopy of the tamarisk trees. Underneath the wisteria are *muguets*, lily of the valley. Their heady perfume mingles with the musk of the box hedge and the pepper of the tamarisks. Hedgerow birds dart here and there, intent on life and nests and song. Even Eida has an expectant air about her and occasionally gallops across her field from breakfast near me to titbits at Juliana's, instead of her usual meandering gait.

Birds are nesting in the virginia creeper, *vigne vierge*, which climbs up a section of the house just next to my bedroom window. A cacophony of sound awakes me each morning, deafening in its intensity. Juliana picks nettles and *mâche* for her salad, along with the wild leeks, *maragane* that are growing between the vines. 'Patriiicia!' she shouts from a distance, waving her hand. 'I'll bring some over for you when I've washed them! *Non!* I'll make the soup instead and bring you that!'

She spends a lot of her day wandering around the environs of Gageac, gathering wild flowers and wild vegetables. Like the turnip that is growing between the vines, their mass of blossom creating great yellow swathes. Or talking to the neighbours. I can hear her voice echoing from the cemetery when she's chatting with Madame Cholet. Or up the road from Pepita's where she's buying eggs and catching up on the latest news of Roger and his drink problem, or at Madame Cazin's collecting her milk.

The cutting off of shoots is finished, and the mowing, and I've almost finished the weeding. It's just as challenging as last year, in spite of my being more adept with the tractor generally. I'm still worried that I've missed some shoots, still worried that the weed-spraying end will hook itself on one of the vines. This fear is not misplaced as it's already happened twice. The second time I needed Gilles's help to disentangle the pipes and spray from a narrow row of vines.

And so to lifting the wires. '*Patriiicia!*' shouts Juliana. She advances towards me, cigarette in her mouth, large pearl earrings in her ears and two diamond rings on her fingers. '*J'arrive!*' and she's on the other side of the row I'm working on, lifting the wire. 'Hand me some crochet hooks,' she says, and we work together, up and down the rows. '*Oh, la!* It does you good to get out into the vines, *n'est-ce pas*? I'd go mad if I had to stay in all day. At least here your worries disappear into the air and are taken away with the wind!'

She tells me of her life in the Midi and, before that, in Italy. She was engaged to a doctor, she says, and one day Yves arrived in her town, Carreres, in search of marble. Her grandfather and father worked in marble, she said, and Yves arrived at the marble factory. 'Of course, he didn't look as he does now!' she continues. 'He was so thin! His wife had thrown him out of the château and he was in a terrible state! *Eh voila!* That's how I came to fall in love with him. Goodbye doctor and hello Yves. And that's how I come to be *Madame la Comtesse! Oui, oui*, Yves is *le Comte* d'Aureil de Palladin,' she continues. '*Mais, je m'en fiche de tout ça!*' I couldn't care less! she says. 'We're all just people, and look at us now!' nodding in the direction of her house.

Michel appears and joins us.

'All the same, I can wear the jewels, *n'est ce pas*?' and she laughs, throwing her head back and running her hand across her brow. '*Ah! Michel! Comment ça va! Qu'il fait chaud!*'

Eida's field is aglow with buttercups and daisies. She is waiting for breakfast and stamps her foot, impatient. Edward, the cat, is in the field with her, stepping delicately between the flowers and following his usual path along the side of the field. Advancing towards the house, he stops occasionally and looks back over his shoulder to make sure he's not being followed. Beyond Eida is the long, low stone building with virginia creeper advancing up it, and beyond that Jean de la Verrie's vines and atelier. It sits just beneath the beautiful ridge of trees, now in leaf, and bathed in the morning sun's rays, bright and fresh.

The vines are in flower again, producing their delicate perfume, and I'm worried about frosts in the night or rain in the day – and the dreaded coulure. '*Eh oui*', says Gilles, '*on ne peut jamais rester tranquille,* you can never rest easy if you're a *vigneron*. You're always at the mercy of nature, *Milledieu!*' You can say that again.

Gilles still has problems with his neck. Or, rather, he has the same type of pain in his neck, but now also at the base of his spine. He knows what it is, he says, it's those screws in the metal plate at the base of his spine sending the pain back up to his neck. They've come loose too, he says. And it's nothing to do with whether or not he got back on the tractor too soon, he shouts, when I suggest that perhaps he should have spent more time recuperating. 'Pamela says the same thing,' he rails, 'and she's wrong too!'

Geoffroy laughs when I ask him about the significance of Madame Cholet's men with straws in their hands. 'Ah yes,' he laughs. 'Sometimes you find people with that problem. You know, they can't work because that piece of straw keeps getting in the way!'

My sweet wine, the Saussignac, tastes very nice. I have racked it again, taken it off its *leas*, and the liquid is now clear and golden, with aromas of honey and apricots. It fermented in the barrels until April. '*Ça mijote*,' says Gilles, it's been simmering away. It was very high in sugar, 28 degrees, when we picked it and therefore very concentrated. And now it's no longer fermenting, but delicious. Instead of lightly resting the cover over the opening of the barrel to allow the gases to escape during fermentation, the bungs have been firmly hammered in to make sure no air can enter. I'm secretly very proud of it and can taste that it's much better than last year.

I'm back on the tractor again, using the spraying machine, *le pulvérisateur*. I've finished spraying at Gageac and am heading over to Monestier, where the ugniblanc grapes and the nine rows of semillon, which give such concentration to the sweet Saussignac, are. The ugniblanc grapes are planted in wide rows, with the semillon in narrow rows. These are my longest rows. They start near the road and rise in a gentle incline halfway along, then slope downwards steeply until almost the end of the row, where there is a short level slipway of about a metre and a half, followed by an even steeper downwards slope leading to a meadow, which also belongs to us.

Driving to the *parcelles* there takes five to ten minutes by car, twenty-five by tractor. As I climb the hill in front of the château leading to Saussignac, then turn left towards Monestier on the tractor, there's time to look around. Among the vines that haven't been mowed are wild tulips, simple and beautiful, with delicate stalks and long yellow cups; sometimes open, mostly closed. *Marguerites* too, large white daisy heads, and violets and *jonquilles*. Along the side of the road are swathes of cowslips and campion. Monsieur Besse waves to me from a distance. We've never been introduced, but we know each other as I always drive this way to Monestier. Sometimes he's mowing between the vines, at other times he's near the road, always with a wide smile, arm lifted, hand waved.

I arrive at my *parcelles*. I always find it nerve-wracking to spray here. The ugniblanc vines are quickly sprayed, being wide rows and only half the length of the semillon. I have to gather up my courage for the narrow, difficult rows of semillon where there's barely room to pass through on the tractor and it's essential to keep a steady, straight course. The tractor seems to speed up on its own after the gentle incline up to the middle of the row and I can't reduce the revs when we start heading downwards because I need them to spray effectively. Thereafter the slope leading into the meadow is even steeper and it's actually rather frightening as the machine seems to take control, like some mechanical monster.

I gather up my courage and set off. Working up the steady incline on the first row and thinking about reaching the end of the row and turning, I decide not to continue down the even steeper slope and into the meadow. It's just too terrifying. I once lost one of the nozzles on the *pulverisateur* before I could reduce the revs. And the return trip up is always as bad, like climbing up a perpendicular wall. There must be a better way.

If I put my foot on the clutch and then the brake at the end of the row, just as I reach the slipway, I'll be stationary. I need-n't reduce the revs at all before I reach the last vines. I can still spray them properly and can reach back to stop the *pulveratiseur* immediately afterwards. Even though it means losing a small amount of spray, it will be worth it as I can then manoeuvre the tractor round on the short slipway, avoiding the steep dip into the meadow.

I hurtle down the first slope, approaching the end of the row. Okay, foot down on clutch. Suddenly, I'm positively racing. My God, brake! As I ram my foot down onto the brake, and take the other off the clutch the tractor rears up, the huge back wheels gripping the soil. The *pulverisateur* is now touching the ground and I'm looking at the sky.

Sheer terror makes me put my foot back on the clutch as the tractor bounces back down to earth, then I stab the brake. It rears up again. The *pulverisateur* throws out a high-pitched, screaming noise as the nozzles hit the ground and the spray jets out. I take my foot off everything and the tractor hits the ground again with force, bouncing up like a small rubber ball, not once but twice. It stays on the ground for a second before hurtling down the steep decline into the meadow. I ram the rev handle down to zero and put my foot on the brake.

Totally shaken and shocked, I bring the tractor to a halt. I can hear the sound of the spray, which continues to disgorge its liquid. With shaking hands, I switch off the taps. The powerful smell of wild lavender and mint, released as the tractor had raced over them and down the hill, mingles with the smell of copper sulphate. There is silence, apart from the screaming sound in my head and my pounding heart.

'*Milledieu!* I've told you about your foot on the clutch!' shouts Gilles. He is passing by as the tractor reaches home, with a much chastened and pale blue figure at the wheel. The blue copper sulphate has dried on me and is starting to sting my skin and itch. Some of the nozzles of the *pulverisateur* are hanging down in tatters from their blue plastic pipes behind the tractor, the rest are next to me on the seat of the tractor.

'You could easily have killed yourself!' Gilles continues 'and who would have found you? It could have been days before anyone would think of looking there!'

This and other thoughts had crossed my mind on the slow journey back. The tractor could easily have fallen on top of me. I could have been killed instantly or, even worse, not been killed instantly, but died slowly. As Gilles rightly pointed out, no one would have known for days.

I dismount shakily from the tractor, take the hose pipe from the courtyard, turn it on and aim it at my arms and face, then decide to go upstairs, wash off the copper sulphate properly and perhaps have a more private weep of relief and gratitude at being alive.

Juliana is bending down and picking something from around the wooden stump of the tree that has been cut down in front of the church. As I open my shutters she shouts '*Bonjour! Attends*, I'll bring you some!'

They are *peupliers*, a type of mushroom that grows on the remains of tree roots. She has also picked *roses de prés*, field mushrooms, from my garden. The *peupliers* are round, very pale and similar to the field mushrooms, but smaller.

'Their taste is stronger and slightly spicy,' Juliana says. She is now in my kitchen and tipping kilos of mushrooms onto the table. 'Delicious! Later we'll have *chanterelles, trompettes de la mort* and *cèpes* to pick!' she continues. She got up especially early, she says, to pick the *peupliers*. 'Otherwise, Michel will steal them all! He has his eye on them too. I saw him looking yesterday evening.'

I feel rather sorry for Michel. He has obviously had unprecedented fiefdom over the mushrooms for years. Until now, that is.

'*Non, non!*' she exclaims, when I mention him. 'Of course I've left him some! *Mais, quand même*, we can have a few, *n'est-ce pas?*' She laughs her raucous laugh and shouts '*Quooqiloo, mes enfants!*' at the dogs. '*J'arrive!* I've something special for you this morning!'

The post has arrived and with it those brown envelopes that I know contain bills or requests for money. I used to leave them for James's visits, but these have become so infrequent I now open them all. Among them is a larger, white envelope. I open it and it's a cheque, from Justerini & Brooks. They had taken one hundred cases of wine, so the cheque was a large one. I look at it for a moment, then rush to the phone to ring James and tell him, then look at it again, put it and me in the car and drive down to the bank to deposit it.

I can't stop smiling. Each time I think about it, a great surge of pleasure overwhelms me, along with relief and hope.

Gilles has repaired the *pulverisateur* and all the nozzles are now replaced and working. He's coming over with me to Monestier, he says, to have a look for himself. He has explained that the front of the tractor is lighter than the back, not just because of the heavy spraying machine behind, but because the huge engine is driven from the back wheels.

We arrive at Monestier and he inspects the slope.

'*Milledieu*, you mean you were attempting to turn the tractor on the ridge without going down the dip?' he shouts. 'Well, you certainly would have killed yourself!' He repeats '*Milledieu*' to himself softly as he looks down into the meadow.

He gets on the tractor and drives down the first row and as he reaches the end, he depresses the rev lever, then turns to switch off the taps of the *pulverisateur*. 'Look, no feet on accelerator, brake or clutch!' he cries, as the tractor careers down the dip and through the wild mint and lavender in the meadow.

'*Quand meme*, I see what you mean about keeping the revs on the slope,' he says and shakes his head.

His voice now sails up from below. 'And I see what you mean about the dip! 'You'll just have to be sure to keep in a straight line and concentrate. It's not very easy, *Milledieu*. And when you come up from here,' he is still in the meadow, 'come up in first and don't be tempted to put your foot on the clutch, because the same thing will happen.'

And indeed, when he tries to ascend, it does. He too instinctively put his foot on the clutch, with the same result.

'*Milledieu*!' he shouts, taking his foot off it rapidly. '*Ce n'est pas très évident*,' it's not so easy.

Michel has brought me mushrooms and some fish, which look like whitebait. 'They're called *les gardonettes*, from the river at Gardonne,' he says. He's been up since 5 am fishing and gathering mushrooms.

I tell him that we ate some of his *peupliers* last week. He laughs, '*Oui, Juliana m'a dit,*' he says. 'Eat these today while they're fresh,' he points at the fish. He glances at the mushrooms. 'Like Juliana, I got there before the others!' He laughs, kisses me goodbye and is gone.

*　　*　　*

'*Il faut frotter le pain avec l'ail*,' says Michel. Gilles, Pamela and I are having a meal at Michel and Monique's. You should rub the bread with a piece of garlic, he is saying, and you'll be healthy as well as enjoying the flavour.

He is barbecuing and we are all sitting at the table on the terrace, just outside the kitchen. We are nibbling on home made pheasant pâté made by Gilles, with some bread and garlic and drinking Clos d'Yvigne rosé.

Michel and Monique's house is small, on the corner of a minor road and a larger road. Neither thoroughfare could be said to be busy. 'Come and have a look at my vines,' he says, 'while the meat is cooking.' We cross the road to the other side, where he has a small garden. In it is a row of vines, merlot and malbec, along with some flowers, a lilac bush and a perfectly worked vegetable patch. He has planted carrots, potatoes, green beans, white haricot beans, peas, green peppers, onions, garlic, leeks, tomatoes and ornamental squash.

We return and I exclaim to Monique: 'When does he find the time to do it all?'

'Oh, la, ask him!' she says. 'If he's not in his garden, he's at the *chasse*, or fishing, or making his wine. Or at work of course.'

Or helping me, I add.

'*Oh la! Ça,ce n'est rien*, it's nothing that. He's often much farther away, aren't you Michel?' She laughs and he reddens.

'At least I know where he is when he comes to you,' she continues. 'It's when he gets up at 3 o'clock and disappears off to Archachon to go fishing, can you believe? Before work!' And she tells us that he was like that before he married her and nothing will change him. He's a man of the soil, a man of the *chasse* and a man of the sea.

'What else is he a man of?' asks Gilles and they collapse in laughter again, with Gilles slapping his thigh, then crying 'Oof!' as his neck and back respond with a stab of pain.

Bruno and Claudie are more or less up to date with their vineyard schedule. Claudie doesn't drive the tractor. Bruno weeds or sprays in the evenings, after work. We're having dinner at my place and they're bringing along some friends, Yan and Sylvie. They've asked me to make another English dish.

They love English food, but I'm running out of ideas. Over the winter, I've exhausted most of the possibilities. Roast beef and Yorkshire pudding, roast lamb with mint sauce, Lancashire hotpot, spare ribs and cauliflower cheese have all been served up. Dumplings were a great hit with them. Not the soggy ones, but the kind that you drop into a lamb stew for the last ten minutes, made with butter, flour and an egg. They have gone home with recipes for rhubarb crumble and real English custard and they both express surprise that the French, including they themselves, have such a bad impression of English cuisine, when it's obviously just as good as the French. All except for the mint sauce with the lamb, they say, as the vinegar in the sauce does not do any favours to a good bottle of red.

Yan is a broker and has worked in London, in the City. He and Sylvie live in Paris, where he now works. He's full of life and anecdotes and interest, relieved to have a few days' break from work. Sylvie, his girlfriend, has the same tinkling laugh as Claudie and all three of them were at school together in the north of France, at Alençon.

Bruno is from Bordeaux, although his father, who was a professional footballer, is Italian. Claudie and he met when she was sixteen and Bruno was seventeen. They were among a group of French students who were spending a month in Ireland on an English language course. They loved Galway, they said, and had English lessons in the morning, spending the rest of the day with the families they lodged with. Bruno had reasonable food at his guest home, he claimed, whilst Claudie's was '*pas terrible*! We didn't learn much English either' he says, 'as we spoke French with each other all the time.'

'*Oui*, and it didn't finish there,' continues Yan 'as they've been together almost ever since and they're still at it!'

And we talk, and eat, and laugh and talk as the sun descends and the moon comes up with a canopy of bright stars.

Chapter 9

WE ARE NOW INTO A HOT JUNE AND THE VINES HAVE FINISHED FLOW-ering. The small green nodules are growing in size. So far, I've sprayed against oidium and mildew as well as an anti-rot solution onto the grapes. Now I worry about possible grape worm, *cicadelle* and *acarioses*. My eyes search for the small, clear, shiny circle on the leaves that signifies the eggs of the grape worm; or for the leaves turning coppery when the *cicadelle*, the smallest of green leafhoppers, descends and sucks all the iron out of them; or the red spots on the leaves produced by dust mites.

I've also become adept at changing the equipment on the back of the tractor, particularly from sprayer to mower, although the weeding machine and the dreaded vine trimmer are another matter. It's time for the trimmer and I know I have to do it before the next spray.

The wine merchants Corney & Barrow ring. Like Justerini & Brooks they are major wine retailers with a fabulous list. They have tasted my rosé and would like to buy some.

My first reaction is that someone is joking, but no. 'Robin Charles, one of our people, tasted it at a neighbour's house where he was having dinner,' says the buyer on the other end of the phone. The neighbour has also, apparently, given Robin Charles a bottle to take to the office. 'We've just tasted it and we like it.

Can you give us a price on the rosé? And can you send us a sample of your red? What are your prices?'

I have no idea what price to charge and I haven't yet got any red in bottles and anyway, right now I can hardly speak.

'Yes, your red's almost ready,' says Bruno as we taste it from the barrels. 'We should think about preparing it for bottling. Bring in an *échantillon* for an analysis and also the Saussignac, for a general *contrôle*.'

In the *chai*, I take a small amount of red wine out of five or six barrels to make up an *échantillon*, then wash out a bottle to do the same with the Saussignac barrels for the sweet wine sample. The barrels are hard to open because I hammered the bungs in forcibly to make sure no air could get in. The third barrel is particularly difficult to open and I tap the edges of the silicone bung with a hammer, then grasp the sides of it with my fingers, pulling and twisting to try to ease it out.

Suddenly, with a massive eruption, the bung bursts out of its hole and wine shoots into the air.

There is nothing I can do to stem the tide as there's no pump to switch off. It seems to hang in the air for a second then crashes down onto my head and shoulders and the floor, smelling of honey, apricots and quince. I stand back and stare at it, at a loss what to do. And it keeps on coming.

The barrel was refermenting and had been for some time. It was only kept under control by the tightly wedged silicon bung. I lose half a barrel, that is to say, one hundred and twelve bottles.

'It happens,' says Bruno when I arrive at the lab. 'Go along to the registrar and tell them about it,' he continues 'and they'll rectify your *déclaration de récolte*.'

'*Oh, la malheureuse!*' exclaims Jean *le régisseur* when I call in on my way back from the lab. 'And you're covered in it!' He laughs and reaches up to his files.

'Have a cup of coffee,' says his wife Annie, 'and would you like a shower?'

Madame Capponi, the bottler's wife, is at my door, file in hand, glasses to the ready. This time it's 150,000 bottles and, *bien sûr*, she's supplying the bottles and, yes, they'll be traditional rather than standard. And, yes please, long corks, superior quality. Back to the printers, then back home again as I've forgotten what alcoholic strength in percentage of alcohol the wine is, then back to the printers again.

I'm sticking to the same quantities of export capsules, namely fifty percent, simply because that's what I did earlier in the year. And, yes, I'm going straight home with the capsules in my car, only stopping briefly at the registry office to depose my *acquis*, the permission to transport the capsules from the FVB to Gageac.

'*La mise en bouteille des vins rouge? Mais alors, ça, c'est sérieux!*' pronounces Jean at the *régie* when I depose my *acquis* and pay my money. He laughs as he stamps the form with a flourish and stashes it among the dozens of files sitting behind the counter.

The results of the red and the Saussignac are fine. The red can now be clarified before bottling and the Saussignac needs to be racked.

It takes two and a half egg whites to clarify each barrel of red wine. The theory is that all the impurities in the wine head to the egg white, which eventually drops to the bottom with them. Racking takes off all the clear juice, leaving the *leas* and egg whites at the bottom. I believe what I've been told so I'm doing it: beat an egg white to frothiness, but not stiffness; with a stick, agitate the wine in the barrel, then add the egg white, stirring all the while. With the remaining egg yolks, either make solid, yellow omelettes, cakes, scones or dumplings, mix with dog food, mix with cat food, add to soups, or give them away.

*   *   *

Some weeks later it's time to empty the red barrels — and it is easy. The clear juice is sucked out through the long, stainless steel pole, through the pipes and the pump and into the vat. Some 95 hectolitres of it is racked without losing a drop, except for the egg whites and *leas* at the bottom of each barrel.

Cleaning the barrels, though, proves a little more labour intensive. I fill them with water, drain them, then fill them again. then drain them thoroughly, then sulphur each one, then re-bung it and finally rub down the outside. It's late, I'm wet, tired and hungry, but it's done.

The film crew are back. They film me mowing. They film me spraying. They film me doing a racking of the Saussignac barrels. As well as filming the vineyard, they're pretty interested in its day-to-day workings. As I now know them better, I find being filmed is easier, as is having them around. In fact, they're rather fun and after the racking of the barrels, they down tools and help me again. Graham and Stuart clean out barrels, while Bernice and Jackie slot sulphur tablets onto the coat hanger tool. They watch and film me sending the Saussignac back into the barrels. I try not to lose any of it and I don't. What's left in the pipes is tipped out and added to the barrels, the silicone bungs are taken out of the bucket of sulphur and water and placed in the holes at the top of the barrels, and a final tap with the hammer finishes the process.

As we all sit on the grass next to the courtyard for a break and to dry off in the sun, we chat about their lives in England and my life here. They all agree it's a pretty good assignment to be sent to the Dordogne to film.

Malcolm, the director, has wandered off down the road towards the *mairie* to see whether there might be some good shots from the top of the hill. He returns to ask whether I know the neighbour opposite the *mairie*. It's Roger.

Malcolm has seen him working in his garden and thinks he'll be excellent filmed as a helper in the *chai*. Can he come along and tap each barrel with the wooden hammer, as we've just done?

I ring him, and am relieved to find that he is relatively sober. He is delighted to be asked and agrees to be in the *chai* in half an hour or so. He arrives, a positive transformation.

Gone is the *bleu de travail*, with one strap hanging down over his stomach. Gone is the liquorice-papered cigarette hanging from his mouth. Gone is his cap and sandals. He is wearing a pink shirt with a multi-coloured woollen, sleeveless jumper, a pair of cream trousers and no socks, but sabots. His hair has been carefully combed and flattened with oil. The perfume of his eau de cologne pervades the *chai* and his flies are undone.

Will this do? he asks, swaying and smiling.

It is time for bottling. This time the bottlers are filmed. It's just as frenetic as the first time. The lorries marry up again into a bottling plant, blocking the entrance to the church. Fork-lift trucks race up and down depositing pallets of empty bottles, and workers swarm around connecting pipes, lifting vat lids and preparing for a day of bottling. Philippe Capponi, the son of the bottler, arrives in the *chai*, inspects the wine and tastes it. '*C'est bon!*' he says, nodding his head in surprise. 'Let's go, men!' he calls, and the conveyor belt jolts into action, bottles clinking, machines jumping into life as the bottles approach them. The bottles are dark green, the wine is a rich red and the labels are white, with a blue and gold edge. They look lovely, sitting in neat rows.

Every so often, Philippe inspects one with an experienced eye and decides the label isn't quite straight, or the bottle isn't quite full. If he decides the latter, the cork is removed and the bottle placed back onto the conveyor belt further up the line to be refilled. If the former, it's plunged into a bucket of water and the label is soaked off and a fresh one applied.

Gilles wanders over to watch. He's in pain again, he says, and it keeps him awake at night. '*Milledieu, ça fait mal!*' it hurts like hell. He's sick of taking pills and he can't lift the stones any more. Even worse, he can't carry his gun for any length of time when it's a hunt day. He'll have to go back to the doctor.

'*Patriiicia!*' shouts Juliana. I can hear her voice clearly, even over the noise of the bottling machines and the fork-lift trucks. She is shouting from her bedroom window. '*Sam est parti!* Sam's escaped again! I can see him from here!'

Sam has taken matters into his own hands for some time now and grabs any opportunity to wander off on a freedom trip. He doesn't go far and has a regular route; up to the cemetery and into the vines, along the road and over to Madame Cazin's, where he's given food and fresh milk. Then back down to Madame de la Verrie's at the château for some attention, then home, but via Pepita's garden and any bitch that's on heat in the area. Pepita is very proud of her garden and Sam's footprints are not appreciated, nor the fact that he demolishes some of the produce on the way.

I'm not too happy about it either. Each time I block a hole in the fence created by Sam, he makes another. The fence is interspersed with large stones, planks of wood and any other object that can stop him escaping.

Luke, on the other hand, has no desire to wander and no obvious longing for a bitch as he remains on the right side of the fence, except occasionally, when he passes through one of the holes left by Sam. Then he stands by the side of the road expecting to be let back in again.

'You should let him come with me on a *chasse*,' says Gilles as I capture Sam and lead him back. 'A bit of training and he'll soon behave.'

*   *   *

Geoffroy is firmly established in his new home. He spends a lot of time going between his house and the château or from his house to mine. He has now retired from Air France and is busy converting part of the château into an apartment for his parents in winter. Or carpeting a room in his house or mowing around the château grounds, which have never looked better. He drops in three or four times a day, for a quick coffee, or to see what I'm doing in the *chai*, or in the vines. He looks tanned and healthy and happy to be in Gageac. I'm happy to have him here too, giving, as he does, friendship and company.

A lorry is expected from Corney & Barrow to collect a hundred cases for England. The driver has rung for directions. He's not far away. Be sure not to come down the hill from Saussignac, I tell him, as you won't be able to turn at the bottom. Half an hour later I see the huge lorry slowly descending the hill in front of the château from the direction of Saussignac. Geoffroy is running up the road towards me. 'He can't get round there!' he exclaims. As he speaks, the lorry is attempting a right turn towards the cemetery, upending a large millstone as it does so, along with some vines.

Someone is knocking at the door. It's a French couple on holiday. 'Can we have a tasting please?'

Momentarily lost for words, I invite them in. I ask them how they know about me. They're in the region, they tell me, and are staying at Sarlat. A friend told them about me and they decided to come and taste. Their friend is the man who collects samples on behalf of the FVB for the *agrément*.

I open a bottle of my rosé and one of the reds and we taste in the kitchen. Mastering the initial feeling of inadequacy, I find I can talk about what I think of the wines and ask them what they think of them. They taste and offer their comments.

How much is it, they want to know? I hadn't given any thought to pricing them – indeed it had never occurred to me that people might turn up at the door wanting to buy from me. For a moment, I'm lost for words, then I remember how much Richard Doughty charged for his so I quote the same.

They buy a case of each. I wave goodbye to them with a sense of achievement.

Following my visit from the French couple, I begin to think about the marketing of the wines in general. I know this is something I'm going to have to find out about, as it's obvious that it's going to be me doing all the selling. Where am I positioning the wine? I'm clear that I want to make the best wine I can, but I haven't given any consideration to marketing. The price I charge for an exported bottle is apparently acceptable to Justerini & Brooks and Corney & Barrow. What do other people charge? How do they sell theirs? And what must I do to sell mine? My head is spinning again and I haven't come to any conclusions, other than to ask Luc and Bruno about it.

It's August and the grapes are setting. It's fascinating to see how the colour change progresses, initially hardly noticeable, then suddenly very obvious. It's at this stage that the ripening begins, with the grapes swelling, softening and eventually completing their cycle. My last spray has just been done. In theory, there is now a month of ripening. I breathe a sigh of relief that I've escaped mildew, oidium and grape worm. In particular, I breathe a sigh of relief now that the last trimming session is over for the year.

In fact, the setting represents a reprieve for all the winemakers in the area. There is a general sense of relief and relaxation. Jean de la Verrie and his wife, Elizabeth, have gone away to see their relations in the north of France. Activity in the vines has ceased, leaving nature to continue its work.

The burnished rays of the sun beat down on them. Cloudless

skies and shimmering heat enhance the soft, graceful curves of the vines and the stillness that comes with intense heat settles on them and the landscape. The dogs no longer rush around the field and it's all they can do to saunter over to Juliana's end of the field for her titbits. The same is true of Eida, who spends her time sheltering from the heat. She lies in the shade given by the small stone house in the middle of her field that was once a baker's oven, or simply out in the open. Edward, the cat, lies sprawled on the tiles of the kitchen floor. If Madame de la Verrie were to place her meringues on the steps of the château, they would be cooked in no time.

Juliana's shutters are opened early in the morning with the usual greeting to wake up the village, then closed for the rest of the day against the heat. Geoffroy has gone to Antibes to spend some time with his friend and check on his *appartement*, and Bruno and Claudie are in the Cevenne at L'Estréchure, where Claudie's family house is. Odile is at Arcachon, where she has a villa by the sea.

'Don't be ridiculous. Of course you can come!' Odile retorts when I say that I can't leave Gageac in case customers turn up. Since my visit from the friends of the FVB man, I've had more visits from people wanting to buy wine. 'Two or three days won't break the bank and you need the holiday.' Which is how I am introduced to Arcachon and the Atlantic coast.

Most of my time there is spent sleeping: in the garden; at the beach; in my bed. The total relaxation that Odile ensured I would have will never be forgotten. We do nothing, other than occasionally saunter to the beach, one minute's walk from her house, for a quick swim. Then back to her sun loungers, positioned in the shade of the maritime pines in her garden. The heat of the afternoon sun makes the pinecones in them crack and hiss. Occasionally, aroused from sleep in the sunlounger, I open my eyes and see, high up in the distant branches, a squirrel leap across from one tree to another.

In the evenings we have Odile's traditional *blanc cassis*, followed by either paella, or oysters, or barbecued fish with melons and salads. They are eaten outside in the garden watching the stars, with the moon casting its mystical light over the garden.

'I'll show you the dunes next time,' she says as I leave. She kisses me goodbye and laughs at the fact that I have come to Arcachon and not seen the famous dunes, even though they were all of four minutes' walk from the house.

'*Ooui, Milledieu! C'était la canicule ici!*' shouts Gilles. It's been so hot here, he says, his dogs can't even be bothered to catch the rabbits who have had to come out of their holes for air. Total armistice has broken out, he says in disgust.

'*Patriiicia! Tu m'as manqué!*' I missed you! shouts Juliana. She has looked after the dogs and the cats whilst I was away. They missed me too, she continues. '*Mais, oh la, laaa, il fait chaud!*' It's hot; it really is! She laughs as she drags her hand over her brow, a hand for once bare of her diamond rings.

It is very hot. The merlot grapes look wonderful, soft, plump and dark purple in colour. The cabernet sauvignon are smaller and harder with slim, more elongated bunches. They hang like elegant jewels. The semillon and muscadelle are swollen, yellow and pale pink. Some are marked with what is the beginning of *botrytis*, with the sauvignon translucent and yellow. Over at Monestier, the ugniblanc is mostly green, although there are a few pale, straw-coloured bunches, enormous and larger than life. Which reminds me of the unsolved problem of pulling them up.

I am having dinner at Bruno and Claudie's on a hot, sultry evening. They both look relaxed and we all look tanned. Yan and Sylvie are there too. L'Estréchure was wonderful, they say, once they got there. And Bruno recounts their journey. He had rung Claudie's uncle, after having consulted the map, to say they were nearly there and to tell them to prepare the aperitifs as it would only take them twenty-five minutes or so.

'We weren't more than twenty kilometres away,' he says.' Only it took us three-and-a-half hours driving over mountains to get there!'

We laugh and have another drink, looking out over the vines gently rising up the ridge, giving softness and beauty to the landscape. A brilliant splash of deep cerise in a glass, the sun sets and a hot, sultry evening washes over us. Stars are beginning to appear, bright and glittering.

'La vie est belle, n'est-ce pas?' says Bruno.

Chantal's baby is born and it's a girl. And she is truly beautiful. Her name is Amy Louise. I am back in England and she is born on the last day of my visit, the 17th of August. She arrived much later than planned and is perfection. As I hold her, I see Chantal twenty-one years ago, in my arms as Amy Louise is now. Chantal gazes at her too, as does Damian, her father and Chantal's boyfriend, all of us in awe at the reality of her.

And I have to go back to France. I can't stay and help Chantal. I can't watch her grow up. It's time to go and catch the plane back.

Damian assures me: 'I'll look after them, don't you worry!'

'Tu m'as manqué,' says Geoffroy. He is looking tanned and healthy and happy. He's just returned from his trip to Antibes. I am washing out vendange *paniers* in preparation for the vendange. I've bought *clayettes*, oblong red bins, which can stack up on top of each other, or be piled compactly away. They're an improvement on the black plastic dustbins and will be easier to manoeuvre when heavy with grapes. As they are shallower than the bins, the grapes

at the bottom won't get so crushed so they'll arrive at the *chai* in a better state.

Vats are scrubbed out, pipes are disinfected and the press is cleaned thoroughly. Not only must I pull up the ugniblanc, I must also consider buying a new press, Bruno has advised. In the meantime, I am in it again, crouched in between the hoops and the large screw running through the middle, scrubbing and disinfecting.

'*Décidément, c'est propre ici.*' pronounces Geoffroy.

There is to be a celebration at the château. Jean de la Verrie's son, Hervé, will celebrate his thirtieth birthday there. The grounds of the château have been mown to perfection by Geoffroy, and Eida's field has been cut too. She looks surprised and bemused at the change in her environment, but not displeased. Geoffroy has invited me as his guest, but it's the evening that James is returning on a late train into Bordeaux and I am picking him up there. 'Mais, tu dois venir!' 'You must come!' he insists. 'Quand même!' he adds, when I say I will.

It is almost midnight as we drive over the hill from Saussignac towards the château. A magical tableau meets our eyes. In the distance is the chateau, lit up from the ground by lights hidden in the grass, the pale grey stone almost translucent. The towers stretch up towards the black sky, which is scattered with sparkling stars. It looks magical; a fairytale château.

It is a hot, sultry evening and all the windows and doors of the château are open. Light radiates from them onto the lawns as I approach the chateau on foot. James doesn't want to go to the party as it's late and he's tired. I have promised Geoffroy that we will at least make an appearance. Beautiful girls in long dresses adorn the staircase, looking graceful and elegant. They are holding champagne glasses and are deep in animated conversation with young men who are standing next to them; others are moving from room to room. They spill out onto the lawns and into the courtyard, the chatter of their voices sprinkled with laughter, echoing through the château and its grounds. I see Geoffroy half-way up the staircase in the hall as I enter the chateau. He is all smiles as he calls me and hurries down to greet me; Hervé, too. All is gaiety and laughter as the normally quiet château is brought to life.

The television programme has been shown. It's called *A French Affair* and is in two parts. It follows the lives of four families who have come to France 'to chase a dream' and examines how reality matches up. I have met the other couples, but only once, at a party given when the film crew had finished.

They have sent me a copy, which arrives the day it is actually shown in England. I watch in fascination and horror as I see myself clambering in and out of vats, wandering through vines with rain soaked hair, attaching *lats*. Do I really look like that? Am I always rushing around like that? The documentary reaches its climax with my vendange of the cabernet sauvignons. We are picking grapes to the gentle sounds of Grieg and I'm smiling as baskets and baskets of dark, purple grapes come in.

Today I am suddenly sunk in pessimism. James is in England and there is no real change in our situation. Chantal is there too, look-ing after a small baby with no-one to help her. The vendange is imminent and I'm about to head into months of frantic activity in the *chai*, requiring huge amounts of energy and stamina.

Then I think of my team of vendangeurs, all of whom have phoned to say they'll be here again as soon as I need them – and my sombre mood lifts as quickly as it came.

I have another recruit from England – Tim, an old friend from Datchet, who is here to lend a hand. Beth, his wife, has dropped him off on her way to Spain and will call back for him in a week's time. Christianne has decided she's not so good at the picking, so has volunteered to do the cooking as well as the serving, to save me some work and give me more time to pick. And she'll do the clearing up too, with the help of Joe.

'*Mais, moi aussi! Je viendrai t'aider!*' I'm coming to help too, exclaims Juliana. 'And so are you Yves!'

Edge is just arrived, with girlfriend. '*Patronne!*' he says, giving me a kiss and hugging me. He seems to have grown even taller, as he towers above me. This year, the Mohican haircut is gone, replaced by a head of curls, but the sartorial style is similar . He is wearing his *Save the Whales* tee-shirt, but this time with long patterned cotton trousers, corded at the waist and ankles, and no shoes. 'Okay, let's go,' he says. 'We're ready to pick. When do we start? Did you get my book of poems? Saw the film, by the way. Great, Boss! Got any more proposals of marriage from it?'

I'd told Edge about some of the bizarre letters I'd received after the programme was shown, including a proposal of marriage. But most of the other letters were full of genuine encouragement and praise.

The *bans de vendanges* have been announced and Jean de la Verrie, Monsieur Cazin, Gilles and almost everyone else has started picking. Gageac is once again a factory.

Madame Cholet passes along the roadside where the sauvignon grapes for the Saussignac are developing noble rot. '*Ah, Madame,*' she pronounces, shaking her head. '*C'est le gâchis!* It's such a waste! Gilles already has his safely in the *chai*,' and she walks on towards the cemetery, hips swinging, bucket over her arm.

'When are we starting?' asks Odile. She's telephoned to say that her neighbour has almost finished and surely we're leaving it a bit late.

'Aren't we ready yet?' asks Pat Chaffurin, my neighbour.

'*Je suis prêt!*' says Roland.

'What would you like me to do?' says Tim.

There's no doubt that I'm stressed. Everyone is busy except me. Bruno arrives twice a week with his *réfractomètre* and measures the sugar content. *'On attend,'* he says.

'But the *viticulteurs* of Gageac have now moved on from their merlot grapes and I'm still waiting,' I say.

'It depends what sort of wine you want to make,' says Bruno calmly.

I wait.

'OK boss,' says Edge. 'What about this wood then?'

Gilles's woodcutter friend Bertrand has delivered my winter's supply. I'd asked if he could sell me some. 'How many *brasses* do you want?' asked Gilles.

What's a *brasse*, I ask?

Gilles doesn't know. *'Milledieu!* A *brasse* is . . . you know! A *brasse*! Don't you know what a *brasse* is?' he shouts, raising his arms upwards and outwards in frustration and to give some indication of the approximate bulk of one.

A dive into the dictionary doesn't help me much as it is listed as a fathom and I don't know what that is either.

Gilles decides I need four of them. They are sitting in two huge piles next to the walnut trees, in one-metre lengths. Two of the *brasse* must be cut in two for the *salon* fire, the other two into four for the kitchen stove. Edge and his girlfriend, Polly, set to work.

They are like automatons, cutting and stacking relentlessly for three days. 'Just give me the energy food,' says Edge. 'Don't forget, I'm in training.'

Polly's obviously in training too. They demolish soups, plates of pasta with beef stews followed by kilos of cheese for lunch, Camembert being the hot favourite.

'I'm vegetarian by conviction, you know.' he says, 'But not at the moment.' and he smiles. I ask if I should stop giving him meat.

'Mmm, let's just play it by ear, shall we? Can I have another helping?'

They both row for a Putney club and Edge rows for Imperial College too – that is, when they're not running marathons, or sundry charity runs, or chopping wood and stacking it.

Tim sets to repairing broken pieces of equipment in the house, or adding extra electricity lights in the kitchen while we wait.

And still we wait. There is no doubt that I'm now more stressed than I was a week ago and the cold sores around my mouth bear witness to this. As do the painful spots on my scalp, diagnosed by Edge and the doctor as *zona*, a form of psoriasis. I wonder why I didn't get this last year and Edge says, 'cos you didn't know what you'd let yourself in for last year, Boss.'

He and Polly have moved on from wood stacking to painting bedrooms and papering ceilings, gathering in the walnuts and burning leaves, interspersed with quick eight-kilometres runs down the hill from Gageac, along the low road towards Gardonne, then up to Saussignac and back home, via the hill opposite the château.

Claudie is also stressed and rings to say that she has eczema behind her ears, and so badly she's worried they're going to drop off. And she's coming to vendange for me because I'll be picking the red before they pick the noble rot.

'Two a row, don't leave your vine until all the grapes are in your basket; don't leave the row to join another until it's finished.' I shout, and we're picking.

My team is grown. Jacques, the husband of Reine, who attends the same English course as Odile and Christianne, is here, along with Claudie, and Jos who is Charles and Laura's son and Joe and Tony, Americans who have recently bought a house in Saussignac. They have come along for the experience, as has Lucy, a Tahitian also living in Saussignac. Pia, my Swedish friend is here, and of course Tim. At least six of my pickers have brought their dogs, all of whom are Labradors, some black, some golden, some of whom bond with each other, attempting to mate instantly, others provoking instant death by growling. And one who decides to chase Edward.

Gilles appears. The harvesting machine in the vines is about to pick all his cabernet sauvignon, which sits next to my merlot *parcelles*. '*Oui*, they've announced rain for tomorrow and I want mine in,' he says. He looks at the sky. 'Mmm . . . It might even rain tonight. I'd hurry up if I were you.'

Relief to have begun is replaced by panic. We're only just starting the merlot and he's picking his cabernet sauvignon. 'Let's go everyone!' I call, and we pick and talk and move up and down the rows systematically, picking luscious, plump merlot.

Periods of quiet are replaced by periods of animation, the newcomers getting to know the others. Charles drives the tractor and trailer, with Edge and Jos emptying heavy *paniers* into the *clayettes* for the pickers and Tim heaving full dustbins onto the trailer. Then it's back to the *chai*. The crusher is operating, and there is noise, manpower and vibration as the grapes surge through the pipe and up into the vat, with stalks spewing out onto the floor. Then back to the vines for more.

'*Patriiicia!*' shouts Juliana, marching towards us in yellow wellies, a black tee-shirt with a gold Chanel logo and red jogging pants. '*Bonjour tout le monde! Oooh! Il y a des hommes! Oh, la, il fait chaud!* Whisky anyone? Gin and tonic?'

She has with her a basket of cold, bottled water. She looks dazzling in large gold and red rhinestone earrings that match her trousers and tee-shirt, her hair swept up with a red comb. She joins my row and picks. '*Tu sais*, tomorrow morning early I'm going in search of *trompettes de la mort* and *cèpes!*' she says. 'It's the right time and I know just where to look for them. *Eh*, Michel, no eavesdropping!' She stops to address Jacques, who she hasn't met before. *'Bonjour, Monsieur!'* He is talking to Roger who is working opposite him. Roger knows his sister, he is telling him. They were in the same school but not in the same class. Jacques turns towards Juliana. She looks at him for a moment, smiles, then turns very slowly towards me, her right hand on her brow, the other on her left hip and with eyes wide and head inclined, *'OOOh, la, qu'il est beau!'* Isn't he handsome! she booms. He is.

It's lunchtime and Charles's makeshift table is in use in the kitchen, as is every dish, knife, fork and glass. Earlier in the day, at 6 am to be precise, I'd borrowed ten chairs from the church opposite. People wander in to lunch while Charles, Edge, Jos and I send the grapes through the crusher and into the vat. Running some juice into a small container and sinking the *hydromètre* into it, the reading is over thirteen in sugar.

There are cheers in the kitchen at the news. 'I dunno what it means, but here's to it,' says Tony, the American.

John calls from Thailand: 'Hi, Mum, how's it going?'

There is a cacophony of sound as people chatter and laugh together. Joe is talking in his slow drawl to Edge's girlfriend, Polly, who is a nurse, about his trip to Canada, 'Well, some people leave their hearts in San Francisco, my dear . . . I left my appendix in Vancouver.'

Jacques is telling Pia how the Italians were not readily accepted in France after the war. 'You know, the Spanish fitted in nicely as they were suitably meek and subjugated, but the Italians, they were proud! When the men came to Mass they were so handsome, with their flat, broad-rimmed hats, their straight backs and black suits. They stole all our women, who fell in love with them instantly! So you can imagine, they weren't very popular with us.'

Back in the vines, and progress is slower for the first hour, while food is digested and muscles readjust to work. We pick solidly through the afternoon, with periods of chatter and silence, laughter and *panier* filling. We still have ten rows to go when it starts to spit with rain. Charles, Jos, Edge and Tim gather up the full *clayettes*, while we pick and pick in a race against time. Claudie pulls on a purple hat, bought in Galway when she first met Bruno. 'How come she always looks so good, even in the rain, when we all look so awful?' complains Pat Chaffurin. And Claudie does.

The following morning in Mass, a fair number of the congregation who have picked for me hobble up the aisle for communion. And the next week they don't go at all. I walk over to the priest who has just arrived to prepare for Mass and explain that one-third of the congregation will be picking rather than praying, even though it's Sunday. And can we please have the chairs again after the service?

'So then, *Patronne*. It's the liquid gold now, is it?' asks Edge. Yes, it is. We pick noble rot, wondrous liquid gold. Carefully, slowly and well. My team, including Claudie and Bruno, and joined by Yan and Sylvie, are dedicated and conscientious. Richard Basque is carrying a *hotte*, as are Edge and Jos.

The grapes are tipped into the *clayettes*; beautiful pale purple and brown grapes, with a delicate white dust. My pickers now think they're beautiful too and occasionally someone shouts 'Oh take a look at this!' or 'A Christmas tree full of bonbons!' Lucy has brought along a small stool and sits at her work. She looks as though she might be weaving, or playing a harp. It's too hard on her back picking with low vines, she says and, anyway, she's comfortable this way. With her long, black hair and colourful straw hat, she looks exotic and incongruous amongst the vines.

Claudie's ears are no longer threatening to drop off, as their vendange is in full swing too. Sitting on the top of a hill, their vines constitute six separate *parcelles*. When I'm not picking here and the *chai* is under control, I go and pick their grapes, with Edge, Polly, Jos, Tim and Charles. They have more muscadelle than me and their vines are also taller. It is easier on the back, says Edge.

Today we are picking at Bruno and Claudie's. Juliana is here too, along with Yves. Not being a picker by heart or nature, but preferring to talk to the *vendangeurs*, Yves picks only periodically, and is shouted at by Juliana. He's not feeling so good today, he says, and thinks he'll go and sit in the car for a while.

We are walking down the *Champs Elysées*, so named by Claudie for its wide, grassy avenue. It leads down into their other *parcelles* of vines and gives a particularly beautiful view of the valley. Juliana has spotted some asparagus growing in a circle, halfway down. Can she come up in the spring, and pick some, she asks? *'Et, où est Yves?'* she says. 'He can't still be up there sitting in the car!' She wanders off to find him.

She returns, white and shaken. 'Patricia! Yves is ill,' she says, urgency in her voice. 'Please come quickly!'

We both climb back up the *Champs Elysees* towards the car, leaving behind the *vendangeurs*. Yves is lying across the back seat, doubled up in pain and moaning incoherently. Juliana is trembling and, by now, crying, standing beside me.

We drive to the hospital, Juliana in the back with Yves, who moans in pain and falls in and out of consciousness. *'Oh, la,* please don't die!' she sobs. With my hand on the horn to warn traffic, we jump the last set of traffic lights, hurtling through them and the gates of the hospital. As we race up to the emergency unit, an ambulance hoots its horn to indicate we can't come through, but we go through anyway because I can't stop and Juliana is sobbing and Yves is dying. Yves is no longer conscious, Juliana is holding him and crying to herself.

We sit in the waiting room, hoping for news. An hour passes

before a doctor comes to tell us that Yves will be operated on immediately for a hernia. There's no point in us waiting here, he says. We should go home and ring up later for news.

We get into the car and drive home in silence, Juliana shrunken and hunched up in the seat next to me. I hold her hand as she cries softly.

In the *chai*, Edge is doing a *remontage* of one of the red vats. 'Thought I'd make a start,' he says. 'Where've you been? What happened?'

It is gone midnight and we are finishing *remontages* and taking temperatures, when Juliana bursts into the *chai*. '*Tout va bien!*'. She's still in wellies, with noble rot stains round her face and nose. The hospital have just rung. Yves is out of danger. '*Personne répond chez toi!* No one answers the telephone around here!'

She's laughing and crying at the same time, large tears sending paths of white through the noble rot stains on her face as she hugs me.

'Beats the nightlife of Putney round here,' says Edge. 'Where's the music?'

The *fête de vendange* was held in the château of Saussignac. Joan and Fred Montanye, the Americans who live in one of the towers there and run a cooking school, offer the *fête* as their contribution to the vendange. We laugh and eat their delicious food and drink, a universal feeling of heartfelt relief that the vendange is over for another year.

That sense of relief to have the grapes harvested and safely in the *chai* is enormous. Even though disasters can easily happen in the *chai*, one feels more in control. Now the vagaries of the weather no longer pose a threat. 'All you've got to do now is a bit of black magic in the *chai*,' says Charles, 'and move it around a bit.'

*　　*　　*

With sadness, I've said goodbye to Edge and Polly. They've been superhuman during their stay. They didn't just do the woodcutting and stacking, and the painting and picking, they also helped pile the skins into the presses and heave around the barrels, to say nothing of their companionship and moral support.

When I'm seeing them off at the station, they claim that it's been good for their training, and kiss me goodbye. I watch them board the train, impossibly huge backpacks filled with wine on their backs. They'll never manage get them back home without breaking their backs, I tell them.

'No sweat. I'll send you some poems when I get back. And watch out for me in the marathon!' Edge shouts as the train moves out of the station. 'And don't work too hard! I'll be back!'

Tim, too, is gone, having been picked up by Beth and whisked back to Datchet and another life.

The Saussignac grape juice is at 28 degrees in sugar. I know it has concentration and complexity. It's clean in taste, already with aromas of pineapple and apricots, and especially honey – and, most important, it has balance. Bruno is at the lab as I deposit *échantillons* of reds, rosé, white and sweet. We all look tired; the lab assistants, Bruno and our first *oenologue* Jean Marc – not to mention the *vignerons* who come and go with their *échantillons*.

The lab is running at maximum with *échantillons* on every available surface. Some have already been analysed and wait to be signed off by either Bruno or Jean Marc. Others are lined up next to one or other of the machines which will measure their volatile acids, carbon dioxide, sulphur dioxide, sugar content or any of the many constituents. Telephones ring as *vignerons* wait to hear if alcoholic fermentations are finished, or *malos* started. The lab assistants work calmly and systematically through the incredible amount of *échantillons* banking up beside each machine.

Jean Marc, my first oenologue, has a vineyard too, with ten

hectares of vines, under the *appellation* of Pécharmant. It was always a separate *appellation* from Bergerac, he's told me, and is known for its good *terroir*; sand, with gravel slopes running along the northern side of the Dordogne valley and only producing red, rather as the Saussignac *appellation* only produces sweet, or Monbazillac, which does the same. Like Bruno, he juggles his work as an *oenologue* with being a *vigneron*.

'*Que c'est bon!*' says Bruno as we taste the Saussignac *échantillon*. Jean Marc tastes it too, and agrees.

I do too, and cannot wipe the smile from my face.

Justerini & Brooks have ordered some red. I sent a sample to them after my first direct clients called and now Hew Blair, their main buyer, calls to place an order. I am beside myself with delight, surprise, pride and relief. He wants to know how the Saussignac is this year and what I think of the harvest in general. And I find I have an opinion on it. He'll be out in the first quarter of the year, he says, and would like to visit the vineyard and taste.

Corney & Barrow, too, have ordered some red, two pallets worth. And I have a new customer — John Davy. He has wine bars in the City of London; indeed, he created the concept. From feeling desperate only a few weeks ago, I now feel elated. I see suddenly that despite my roller-coaster existence here, where I constantly swing from some funds to none and back to some again, from hope to despair and back to hope, I have a real determination to continue.

I've got to pull up the ugniblanc and I'm going to do it this year, along with a small *parcelle* of cabernet sauvignon. Thierry Dhauliac has said he'll come over with his father's huge tractor and chains and help. I'm driven on by events, it seems to me. Luc is insisting, Bruno agrees and it's true that I can't make a decent dry white with it. It's therefore a waste of time picking it, a waste of effort tending it and a waste of money spraying it.

Which will mean that I won't have any dry white.

I consider the possibilities. I could take some of the semillon from the sweet wine *parcelles* and the small amount of sauvignon I've got to make some. But I'm fast becoming obsessed with the sweet and I know the semillon I use for it gives great concentration, so it's not really an option.

A Corney & Barrow lorry is arriving to pick up a consignment of red. It doesn't try to come down the hill from Saussignac, but arrives by the sensible route, up the other hill and past the château. It's another monster of a lorry. I watch it advance along the road past the far side of Eida's field. She watches too.

Instead of turning left towards the church and me, he continues straight on, past the cemetery and up towards Jean de la Verrie's château, Jean Brun. The driver gesticulates from a distance. He thinks the turning curve is too narrow, because of the mounted cross in the middle of the road. He's going to find somewhere to turn and come in from the right hand side.

Geoffroy is here again and walking up towards me. 'It's true, that cross is a nuisance,' he says. He's back in Gageac, but only for a week. He and his friend are off to Annecy. Have I been there?

No, I haven't. In fact, I haven't been anywhere really, apart from Bergerac and the lab or Ste Foy la Grande and Leclerc, the large supermarket where I buy food. Charles and Laura, who are now living in France, their house being finished, spend a lot of time visiting different places and know much more of France than I do.

'Oh Annecy is wonderful,' says Geoffroy. It is one of his favourite places in France, but then, there are so many. 'However, at the end of the day, you can't beat the Dordogne and Perigord, can you?' he says. 'Do you know the story of how there came to be so many châteaux in the Dordogne?' he asks. An angel was passing by.

Like the one who sips my wine, I reply.

She had an apron full of castles, he says, and as she flew over the Loire, she dropped one here and another one there. The same with the Limousin, and the Lot. When she got to the Dordogne it was so beautiful and she was so tired, she emptied her entire apron of châteaux over it and flew back to heaven.

We agree that Perigord and the Dordogne is really very beautiful. It's a nice story anyway.

As we wait for the lorry, we gaze down the road towards the cross and I ask Geoffroy why it is there. There's also one at the crossroads to Jean Brun, though it's a much simpler one, in wood. And there is the dilapidated one next to Madame Cholet's, smaller than the tall, handsome one in front of us, but more decorative, beautifully fashioned in wrought iron with vine leaves and grapes, interwoven with a chalice and chasuble and an effigy of the Virgin and Child.

Geoffroy's not sure, but thinks they are mission crosses, dedicated to a Saint's day or a feast, or simply to commemorate an event in the village.

The lorry is taking a long time to turn round. We look in the direction of the noise and see its stately approach from the cemetery, turning perfectly into the road and up towards us and the waiting cartons, with the telephone wires of Gageac festooned over its body and a wooden telegraph pole following in its wake. Jean de la Verrie is running behind it, arms waving, shouting to the driver to stop.

Chapter 10

'*PATRICIA! BONJOUR!*' IT'S MADAME QUEYROU, IN THE VINES SOME distance from me. We are both taking off the links from the wires in preparation for pruning. 'You know, Patricia, that Gilles is thinking of selling his vineyard?'

I don't.

'*Eh, alors*, we thought you might like to know. We don't want any foreigners to have it! You should buy it!'

I'm speechless. I am intensely pleased that I'm no longer considered a foreigner in the village and have moved up the ranks and into the honoured state of a Gageaçoise, but intensely alarmed that Gilles is selling.

Why is he selling? I ask.

'*Oh, la*, his neck and back. It's getting him down and it looks as though he'll have to have another operation.'

'Oui, it's true,' says Gilles when I arrive in his kitchen shortly after. 'I was planning to come over and tell you this morning. It's not that I can't do most of the physical work, even the tractor. It's those stones, and it's not getting any easier and I'm not getting any younger.'

He's only fifty-three and he doesn't have to heave the stones out of the ground. They're part of the *terroir*, I tell him. I don't do it and my land still produces grapes. And it's the same land.

'*Oui*, but you don't plough the land between the rows,' he replies. 'And while we're about it, I don't know how you have any grapes at all, not ploughing fertiliser in! I have to gather the stones up because they break my machines. I wouldn't have any grapes if I didn't work the land.'

He continues. 'It's not just vines, there's also the wheat and the sunflowers. And I don't want to end up a cripple.'

I feel terrible for him.

He laughs, looking at my face. '*Milledieu!* It's not that bad! I want to enjoy my life too! Have a few holidays!' Then he says: 'Why don't you buy it? It's rare that vines next to your own come up for sale, and I'd like you to have it. I could even do a couple of days' work for you – part-time. It would keep me occupied and pay my social charges without overstraining my back . . . And I could keep you on the right track!' he adds.

Walking back to my house, my mind is racing. I can't possibly buy it. I haven't got any money. It would be folly. And why give myself five times more work? I can hardly manage what I've got. But he does have sauvignon rather than my awful ugniblanc, and some muscadelle, plus the merlot; and he's got a lot of semillon, all of them overproduced but with great potential. And there's the cabernet sauvignon. And his *terroir* is the same as mine.

Suddenly, I want them.

How can I find the money? We don't have any and it's a constant source of worry. It's true that I now have some real customers and sales, which at least confirms that I'm doing something right. And it does now bring in some income. However, I'm about to pull up the ugniblanc and some of the cabernet sauvignon, which will mean less red and no white. The white is not such a problem, as I don't sell it to any trade customers.

In any event, I either have to have more vines, or give up.

Pulled up short by the prospect, I stand and look around at the vines surrounding me – my vines. Giving up is not an option. I haven't got this far to just give up. I will do it.

'*Tu es complètement folle!*' says Odile.

'You should!' says Bruno.

'You must!' says Luc.

'*Tu as un sacré terroir*, you have great terroir. It's already yours,' says Jean Marc.

I go to see Gilles. Don't sell to anyone else, I say, but give me three months to find the money.

He says he has already put the vineyard in the hands of an agent, but won't do anything until I come back to him.

I'm going to try and find fifty people to each buy an advance purchase of wine from me, so that I can buy Gilles's vineyard. Each one will pay £5000 and buy one hundred cases, but in advance, receiving ten cases a year of their choice, delivered to their door. They will become bondholders.

One friend in England is already interested – Malcolm Farrer Brown, a solicitor. He has also offered to help with the legal side. 'If you're going to do this, let's do it properly' he had pronounced.

'I feel sure you'll find them,' says Nick Ryman, a local wine-maker of almost legendary reputation. 'I did.'

I met Nick Ryman some time before, having heard much about him. In Bergerac he was widely held to have almost singlehand-edly improved the region's wines with his innovative approach to wine-making when he arrived here twenty or so years ago. The consistent quality of his wines, and his determination always to improve on what had gone before had lifted the *appellation* and encouraged local winemakers to improve their standards.

He had a bondholder system that had been running for fifteen years, he said. Indeed he had three hundred and eighty-five bondholders. I didn't want anything like that many. Now in his late sixties, Nick was retired, although still living at Château Jaubertie with his partner, Eve. He had given me encouragement from our first meeting, and he had often said if there was anything he could do to help, I had only to ask. And now he is offering his help again.

My mood swings from enthusiasm and optimism to doom and pessimism. A number of my friends in England have already agreed to be bondholders, not even knowing what the deal will actually be. In black moments I can't imagine finding more than three or four more.

'Patriiicia! Tu es là? Regarde moi ça!' shouts Juliana through the kitchen door, back from picking mushrooms She is wearing a yellow hat to match her wellies and is carrying a large stick and a huge plastic bag. In it are *cèpes*, the horribly expensive mushrooms and speciality of the region, along with *trompettes de la mort*, small black mushrooms with long stalks and serrated edges. She's been out since 5.30 this morning, she says, the best time to find them, before the rest of the world comes to pick them.

'You can meet half the Dordogne up there!' she grumbles, disgusted that others should have stumbled upon her mushroom paradise. She tips some out onto the table: 'Have some,' she says. 'Are you going to buy Gilles's vineyard?'

She's heard from the neighbours, she says, and, of course, getting anything out of Gilles is impossible, so there's no better person to ask than me.

I don't know, I say, as I have to find the money first. *'Oh, la! Façile!'* she says, banging her hand down on the table and looking at me seriously. 'Ask at the shop! They'll give you some; you can buy it by the kilo!' and her laughter fills the room.

Chère Patronne says Edge's letter. *You should buy it. More rigorous training for me. Love, Your ouvrier favori.*

PS I'll be sending you my poem book in the next week or so. It's to raise money for the blind. If you like it, send some money to the Blind Society. Here's a sample.

> *Blown Away by a Fan*
> *One day*
> *A hot day*
> *I got some imaginary fan mail*
> *But unfortunately*
> *It got blown away by a fan.*
> *I was really blown away*
> *Well, not really*
> *Imaginarily.*

Edge

I've got to buy a transpallet and a fork-lift truck. Jean de la Verrie continues to lend me his transpallet at a moment's notice, but piling cartons on to lorries continues to be a nightmare. In fact, it's getting worse as orders increase. For the moment, Juliana or Michel or Geoffroy, when he's here, help to hand the cartons up to the driver. But it's crazy that I have to transport them down to the road carton by carton when they're already sitting on pallets to start with, only to re-stack them onto the lorry.

Which is what we are doing right now. The lorry is here and Geoffroy and I are handing them up to the driver. This time, while turning in towards the church, the lorry has knocked the base of the cross at the bottom of the road, which has buckled. The cross is now leaning at a gentle angle, pointing towards the cemetery.

'Il faut que je te donne un morceau de terrain,' maybe we ought to give you a bit of Eida's field to widen the road, says Geoffroy.

Morceau by *morceau* I'm doing a fairly good job of demolishing Gageac, I tell him.

He laughs. He'll talk to Jean and his parents, he says. They could give it to the commune, who could then widen the road.

223

* * *

'*Bonjour, chère Madame.*' I've gone to see Monsieur Roy at Foyenne Motoculture. I want to know how much fork-lift trucks and transpallets cost. Transpallets cost sixty times less than fork-lift trucks. I buy a transpallet.

I'm meeting with Malcolm Farrer Brown and an accountant friend, Christopher Lowe, to work out the deal I'll be offering my potential advance purchasers. I have with me my *déclaration de récolte*, a list of my stock and a rough idea of how much red grape and white grape wine Gilles produces. I need to work out how much potential harvest I'm going to have if I succeed in buying his vineyard. The day passes in questions, calculations, possibilities, practicalities and realities. We discuss types of wine, types of expenses, rates of interest, tax and duty charges, transport and delivery rates, possible disastrous harvests, legal implications and personal injury – and much more besides.

'We'll give the Clos d'Yvigne purchase scheme a launch at our house,' suggest Bob and Marie, friends from London. They and other friends in England are enthusiastic and supportive, also providing names and addresses of friends of theirs who they think might be interested in buying into the scheme. 'So hurry up and get it done,' they say.

'It's a great deal for them. You won't have a problem,' says Nick Ryman. He, too, has suggested some names.

Gilles is noticeable by his absence. He no longer calls in. '*Oui*, because I don't want you to feel harassed,' he explains, when I go round to see him.

I am harassed anyway, I tell him. And I've come to say that I definitely want to buy it. I explain that I'm not yet sure I can find the money, but have decided to try.

'*Milledieu, si tu as decidé . . . !* If you've decided . . .' he smiles.

He tells me that *la Safer*, a state-run agricultural organisation, should be informed. 'In any case,' he continues, 'it's cheaper to pass through them.'

I'm not sure what *la Safer* is, but I agree.

Buying agricultural land requires their agreement, Gilles explains. They are a government-backed organisation set up to ensure that farms aren't broken up and the land sold off. They have the right of pre-emption on all agricultural land.

'You mean they might buy it?' I ask him, alarmed. I now feel I have proprietorial rights on the land.

'Oui, c'est possible. And if they find a young *vigneron* who also wants to buy it, they'll buy it and re-sell it to him. So hurry up and get it done.'

Monsieur Dimec of *la Safer* knocks on my door, armed with a thick file and a brief case. And indeed, they do want to buy it. However, they will buy it in order to resell it to me, he tells me. With fewer supplementary charges to *notaires* and the like. The price Gilles is asking is a good price, correct for him and also for me. As far as the house is concerned, he can't comment. This isn't a problem as Gilles and I have agreed that he and Pamela will stay in the house and I'll simply buy his vines.

However, Monsieur Dimec points out, they have already had an enquiry from a *vigneron* who is seriously interested.

My heart misses a beat. What does it mean if someone else is interested?

'We take into account their circumstances,' Monsieur Dimec continues, 'and their needs.'

But I need it and, furthermore, I want it! I want to shout.

The printed brochure for my potential advance purchasers is ready. We have worked and worked on it, Malcolm Farrer Brown as the legal eagle, Christopher Lowe as the numbers man and me as the provider of the text. Nick has given advice whenever I've asked. The brochure looks professional, simple and clear and now it only remains to send them out. I sit up late into the night, sign-ing the accompanying letters and addressing envelopes to the people on my list.

'Il faut couper à mort. You have to cut them dead. I'll go over tomor-row and make a start,' says Michel. He is going to start pulling up the ugniblanc and we are standing outside with Gilles. The pruning is done, and most of the piquet and wires are replaced.

Gilles has also finished pruning on his land. *'Milledieu!* My arm!' he complains. 'It's not the steel plates, just the pruning . . . and the gun,' he continues.

Michel and he were at the *chasse* yesterday. Most of the day seems to have been spent having a long lunch at Gilles's house with all the other *chasseurs.* It was some sort of celebration and I gave them some cases of rosé.

'We had a good day, didn't we Gilles?' says Michel.

'Oui, oui!' Gilles agrees, and they recount their day.

It's now one week since I sent out the brochures. I have stopped calculating what day and what hour of the day people will have received my package. I'm trying not to think about it too much. Particularly as I'm about to pull up the ugniblanc.

'Patriiiicia! Sam s'est echappé!' shouts Juliana from her upstairs window. Sam is wandering down the road on his way to Madame Cazin as the postman's van advances. The postman leaps out and grabs him by the collar.

'Eh, oui,' says Gilles. *'Il cherche les femmes,'* he's looking for women again.

You bad boy, I say as he wags his tail in agreement.

'I've got letters for you' says the postman. 'From England. A lot of them.'

Gilles and Michel look at me quizzically as I tear one of them open. It contains my yellow order form, along with a cheque and a letter. 'Delighted to take an advance purchase order.' Another says: 'Hope I'm the first on the list'. There were fifteen of them. I hug both Michel and Gilles in excitement, disbelief and happiness.

Over the next week or so, letter after letter arrives, and I have my quota of bondholders. I feel humbled by it, overwhelmed with awe and gratitude.

I go to see Gilles and make my offer, explaining that I won't be able to increase it if the other *vigneron* who is interested offers more. La Safer rings to say that the other *vigneron* is definitely interested and will be making an offer to Gilles. My heart is in my boots.

'Look, we know you and we don't know him, and we want you to have it.' Gilles and Pamela are sitting in my kitchen.

Gilles continues. 'His offer includes the house and it's a better offer than yours.'

'But I thought you wanted to keep your house,' I say, feeling demoralised and, by now, resigned to failure.

'We did,' Gilles says. 'But having thought about it, it will be better all round to buy another. We know you can't offer more. What we thought was that we'd sell the property and the vines to you for the same price as he's offering, but spread the difference over three years. Interest free.'

I look at them in disbelief. They're smiling at me. I accept.

'*C'est fait.*' It's done, says Gilles. I'm smiling too.

'I'll help you, Madame!' says Madame Cholet. She is holding in her hand a spray of mimosa she has brought over for me from her tree, which is in blossom. She's always worked in the vines, she says, ever since they bought the property. So it will be second nature. 'And anyway,' she continues, 'you're going to need some help!' And she heads off towards the cemetery with her young girl's walk and a bucket on her arm. She turns and adds 'I'm glad you're buying it, Madame.'

I've decided I'm not going to plough between the rows because I need to restrict the growth of the vines.

'*Ooui*, you won't have any harvest!' shouts Gilles as we sit in my kitchen drinking a coffee.

And I'm going to let the grass grow between the rows.

'*Milledieu!* The vines won't have any food!'

And I'm not having double *lats* any more and I'm not putting down any fertilizer.

'*Oh, la, c'est foutu!* They've had it!' he exclaims.

I thought I'd better tell him and get it over with in one go.

'*Oui,*' he says two days later. 'It's true, you have a different method from me.'

We are looking at his *chai*. Jean Marc has said that it's always a mistake to work from two *chais*, even if they're close together. Pumps and pipes will always be in the wrong place, or vats, or other equipment. Bruno and Thierry have said the same. Thierry Dhauliac does, in fact, have two *chais* and says it's nothing but *emmerdant*, a real pain.

Looking at Gilles's *chai* and calculating how much wine I might have, I decide to use mine instead. Which means contacting Serge again in order to enlarge it.

And leaping from four-and-a-half hectares to twenty-one is no mean feat either. Gilles wants to work two days a week, which I'm delighted about. Michel says he'll come and lend a hand too, especially as he'll soon be giving up working as *cantonnier* for the commune.

Serge is back. We have decided to double the capacity of the existing *chai*. But it has to be finished by the vendange, I tell him.

'Mmm . . . by the vendange . . .' he repeats slowly and we laugh, remembering how he always repeated whatever I said to him in the early days, 'Mmm . . . a duvet in the atelier . . . , . . . Mmm . . . hang my coat on this ash tray . . .'

Dear Patronne, So you got it. More liquid gold, I expect, more training. Got a new turbo-charged babe, Kim. She's in for some training and she already knows about the grape skins on your head and the sticky stuff.

> Poem (One Year Old)
> Edge,
> Nine,
> Tee,
> Nine.

Edge '98
Love, your ouvrier favori.

'*Mais alors, c'est une affaire qui marche!*' says Geoffroy, it's a serious business now. Serge is hard at work and the breeze blocks are going up rapidly. Planning permission took longer than usual as the house and *chai* are close to the château, which is classed as a historic monument. All building work within a five-hundred-metre radius has to be approved by the architects of *Bâtiments de France*, the equivalent of English Heritage. We've agreed to conform to all their regulations, including that the new building will be clad in old stone, like the house.

I think about Geoffroy's comment. It's not true that it's a serious business. It's true that it's beginning to look like a proper enterprise. And it's true also that I have some real customers in the form of Justerini & Brooks and Corney & Barrow. Indeed, J&B are now taking red as well as the sweet, and are taking more and more, but it's all been very much subsistence winemaking so far, and it still is.

One thing is clear to me. Although I never chose to do what

I'm now doing, and did it because I had to, I now do it because I really want to. When did I decide that?

Probably when I was driving through the vines mowing and noticed grape worm eggs on the leaf of one of the vines. I remember seeing them, recognising them and being outraged that one should have laid its eggs on my leaf. Then, realising that I had identified them without any help, I suddenly felt proprietorial, and determined.

Or when I tasted the sweet wine in the barrel and recognised what it actually was, thanks to Jean Marc's *dégustation* course.

Or when I decided to rack one of the red vats because I smelt reduction in the wine. And, knowing that aeration would remove it, did one without seeking confirmation from Bruno.

And I knew it absolutely the moment Gilles suggested I buy his vines and I knew I had to have them.

Still I feel overawed at the purchase of Gilles's vines. Here is the beginning of another roller-coaster. What have I let myself in for? I'm going to have five times as much work and five times as much wine. I have no idea if anyone's going to want the wine when I've made it. Nor what's it going to taste like? I haven't ever tasted any of Gilles's wine. And what about buying more barrels? Or the sprays? Or the bottling?

I decide not to think about selling the wine for the moment. Or the taste of it, or the barrels or the enormity of the task ahead of me. There's more than enough to think about with the building of the new *chai*. And in the fact that I'm now thinking in what seem to be industrial quantities of juice compared with what I've been making so far. And there's the dry white . . .

Gilles's vineyard has sixteen hectares of land, twelve hectares in vines and the rest in agricultural use. We walk round and he points out areas where he plants wheat or sunflowers. I decide I am definitely not going to complicate my life with worries about wheat and sunflower crops. No doubt there'll be yet another set of forms, with accompanying rules and regulations. To say nothing of the steep slope I see before me which some crop will sit on, requiring a spray or two.

The delectable baby Amy Louise is here with Chantal, who's expecting another baby. Amy is beautiful and the very image of her mother. Odile coos at her, Christianne coos at her, Geoffroy coos at her. Even Gilles coos at her. But most of all I do. She's just started walking. Sam and Luke inspect, as do the cats. She's taken to say hello to Eida. Juliana wants to steal her away. All too soon her stay is over and they're gone.

Sam and Luke look for her the next morning and spend the next few days in the doldrums. Inspecting the house in vain for signs of her, they collapse on the floor in a flurry of legs, paws, heads and drama, letting out long sighs periodically. Sam restores normality by planning and achieving a great escape, leaving yet another hole in the fence.

Bruno has decided to give up working at the lab as an *oenologue* to concentrate full time on his vineyard. There's more than enough work for him there, he says, and he knows he has some great *terroir*, which he wants to exploit to its full potential. 'You have good *terroir* too,' he continues.

We taste from the barrels in their *chai*. Their wine is delicious. Full of concentration and complexity, it is intense with aromas of ripe fruits, apricots and honey. They're presenting it to the *Guide Hachette*, the wine buyer's bible in France, for tasting and judging. I'm doing the same with my Saussignac. 'You need to think about your dry white this year,' Bruno says. 'You could make something really good.' I do think about it; a lot.

*   *   *

It's July and I'm back with Monsieur Roy to buy a cooling machine for the dry white vinification. '*Ah, bon.* What size?' Here we go again. Monsieur Roy shows me a number of them. I leave, having signed up for one that he promises me will change my life. It does, as I find out when I make the white wine.

Gilles is looking for somewhere else in the area to live, but hasn't found anything yet. He can stay just as long as he likes in his house, I tell him. In fact, the longer the better as I have no plans for it. It's rather a beautiful house from the outside. Long and low, with a steep roof housing four windows, it's faced with grey cement. But underneath are beautiful white stones, Gilles has told me, just like the adjoining *chai*. Apart from the kitchen, I haven't seen the inside of the house. How odd, I think – in England we wouldn't dream of buying a house without seeing the interior. Still, its no more odd buying a house you haven't seen the inside of than buying vines when you haven't tasted what they produce.

At the bottom of the track between his house and mine is Madame Cholet's house, smaller and a little ramshackled, but also made in beautiful stone. She has a vegetable patch behind it, and nearby is her mimosa tree and the dilapidated cross. In front of the house is a large mulberry tree under which she sits in an armchair each afternoon, with the local newspaper and her cat.

When her husband was alive, they lived in Gilles's house. They had cows and grew maize and beetroot as well as vines. She moved into her present one when Gilles took over the vineyard. From the age of twelve until her marriage, she had worked as a domestic in a large, *bourgeoise* house in Bordeaux. Her wages were her board and lodge. '*Oui*, Madame. I have always worked' she says with fierce dignity.

Gilles oversees my work as I mow up and down the rows outside his house on my tractor. 'Non! Get in closer to the vines!' he shouts, even as I brush the leaves of the vines with the huge wheels as I drive alongside the rows. 'And you have to mow three times up and down these! Don't think you can do it in two because it won't be well done!' Gilles's vines are '*vignes larges*', wide rows. They seem almost as wide as Bruno and Claudie's *Champs Elysées*. '*Oooui*, I told you it'd be hard work! Much easier to work the rows with the plough!' he shouts. 'And don't forget, there's the tying up of these young plants afterwards!'

Madame Cholet looks up from her newspaper, looks at the young plants and smiles at me meaningfully. 'Maybe I'll help you with those,' she says, then turns back to the paper.

Peggy, Odile's daughter, has given birth to a son, Milan. She and her boyfriend came back to France for the birth, as they wanted the baby to be born here. Odile is as overwhelmed and as captivated with her grandson as I am with Amy Louise. Shortly afterwards, Christianne and Richard's daughter also gives birth, to a son, Hugo. And Carole, Christianne's second daughter and faithful vendangeuse is also expecting a baby. 'Potential vendangeurs, all of them!' pronounces Odile.

'*Decidément!* A positive population explosion around here,' says Geoffroy at dinner that night, laughing.

We are eating at the Moulin de Malfourat, a restaurant at the top of the hill in Monbazillac. Run by Paul and Marie Rouger, the food is delicious. Marie is the chef. Blonde, tall and elegant, she visits each table during the evening to check that all is well. Paul is the perfect restaurateur; discreet, overseeing the restaurant as a whole, taking orders, and managing his waiters. I know them well, as I have held my *fête de vendange* there. The restaurant has a panoramic view of the valley, including Bergerac and the small villages to the left and right.

Geoffroy and I look out over the valley as we catch up on each

233

other's news over Marie's delicious food. We've asked her to make us whatever she likes, '*l'humeur du chef*'. The result is wonderful. A sweet, Monbazillac sauce accompanying succulent hot foie gras that melts in the mouth is followed by *sandre*, a local fish from the Dordogne, cooked to perfection with a tangy mix of duck stock sauce and sultanas. We finish with a sweet, luxurious *flambée* of apple tart, all presented with flair.

Geoffroy will be around for the next month or so. He's building another bathroom in his house, and another bedroom, he says, as he wants to spend much more time here. He has almost finished converting part of one of the towers of the château into an *appartement* for his parents, and is now charging on with the new bathroom in his own house.

When he's not mowing around the château, he's cutting wood. Or painting rooms. Or transporting pieces of furniture from the château to his house. Or vice versa, interspersed with quick visits up to say hello, always with a smile or an anecdote or a complaint when some piece of work isn't going according to plan. His friend is coming to Gageac for a weekend, he says excitedly on one of his trips up to me. 'It's not definite, but it looks as though he might have the Friday and Monday off work, which would make the trip from Lyon worthwhile.'

My Saussignac tastes delicious. Jean Marc is with me in the *chai*. '*Je te récupere!*' he's got back his client, he says, laughing. Bruno has now given up his job at the CIVRB to concentrate on his vineyard. Jean Marc no longer works at the CIVRB either. He now has his own business and is only taking the clients that he chooses. The list of *vignerons* who want his services is long and I'm relieved and happy to be among those who have been chosen.

We discuss the forthcoming vendange and the type of dry white I'm hoping to make. I'm beginning to feel distinctly worried about the increased quantities that are going to come into the *chai*. In fact, I'm more than a little worried about the new *chai*. Serge is working hard on it every day, but it doesn't look anywhere near finished. We haven't yet got a cement floor, the roof isn't on and we still have only breeze blocks for walls. And even the walls aren't entirely constructed. The *pressoir* is sitting outside covered in dust and we're already into July.

'*Eh voilà*. Take the rubber cord, choose the straightest shoot and tie it onto the bamboo stick, one at ground level and one fifty centimetres further up. Then move on to the next one. But don't tie it too tight as it's going to grow fatter — and longer, *bien sûr*. But we'll talk about that later. And if I were you, I'd wear a hat!' Gilles is introducing me to tying in the young plants in front of his house.

All vines in France are grafted onto American rootstock since the problem with *phylloxera*, a bug that almost destroyed the vineyards of Europe a century ago. From the rootstock, grow two or three shoots, each around twelve inches in length. One of them will become the trunk of the vine, and that is what I'm tying in. Squatting down to ground level (on uneven ground), I choose my shoot, tie it in carefully at two levels, cut the cord, cut off the shoots not used, stand up and move to the next vine. Then squat down again and repeat. Multiply this by half a hectare.

The heat beats down on my back, legs and neck. By lunch time I'm dehydrated with stiff legs and an aching back. In fact, I can hardly walk.

'I think I'll give you a hand later on,' says Madame Cholet, chuckling.

'It's hard, isn't it?' laughs Gilles. 'It doesn't matter how fit you are, it's the different muscles you're using.' Even he found it diffi-cult, he says. It was one of his least favourite jobs. 'You have to work through the pain', he laughs again, watching my attempts to straighten up and walk. 'By the third day, you'll be fine. I'd help you, but my back's giving me hell!'

And it's plain that it is. For the last month or so he's complained about it. He has difficulty sleeping and it's getting him down. '*Milledieu!* Don't you start!' he shouts when I suggest he should see the specialist again.

Pia, my Swedish friend and *vendangeuse extraordinaire,* arrives to help me, looking very professional in boots, tracksuit bottoms, long-sleeved tee-shirt to guard against the sun and a hat.

'I've come to help, Patricia. I'd love to!' she says.

I explain that it's not for the faint hearted.

'But I'm fit; I walk such a lot and ride. Really, it'll be fine.'

I think of yesterday's effort and the pain levels in my muscles. I can walk again this morning, but they are still complaining strongly.

Madame Cholet appears and gives us each a small apron with a pocket in the front. 'To hold your roll of rubber cord,' she says. 'You'll need them.'

One hour later and I look back down the row. I see Pia's discarded tracksuit bottoms and tee-shirt abandoned half way along the row. She is soldiering on, but in difficulty. She is now sitting on the ground to tie in rather than squatting, and only moving on to the next vine by willpower and determination, both of which are obviously waning. When we stop for lunch, she moves slowly towards her car in a crab-like fashion. She might not be back after lunch, she says. She isn't.

A letter arrives from the *Guide Hachette. Chère Madame*, it says, could I furnish them with an example of one of my Saussignac labels? They would ask me not to mention this to anybody until September when the book is published. I ring Bruno immediately. What does it mean?

'It means that you've won a *Coup de Coeur* for the wine,' he says. And what's a *Coup de Coeur*? I want to know.

It's the icing on the cake. Over 30,000 wines from the different *appellations* in France are tasted blind for the *Guide Hachette*, with only 8000 or so included. Out of those, a mention is good, a star even better, two stars remarkable, three exceptional and a *Coup de Coeur* is all the above with the icing.

We go out to dinner to celebrate six stars and some icing. Bruno and Claudie have one too for the wine from their vineyard, Tirecul la Gravière.

Chapter 11

THE BREEZE BLOCK WALLS ARE NOW FINISHED IN THE *CHAI* EXTENsion, along with the insulation and most of the wooden structure of the roof. The electrician is here to start wiring. Suddenly, things are looking more professional. He wants to know where the *pressoir* will be in relation to the crusher. How many sockets will be in use at any one time? Where will the vats sit? Where in the *chai* will the vendange come in? Am I pressing outside or inside? Where will it be most practical to have outside lights? Am I vendanging at night?

'Are you ready for them? We're coming over now.' Gilles has telephoned. He and Michel are transporting the free-standing vats from Gilles's *chai* over to mine. I'm going to need every available receptacle, says Gilles and I believe him. Each time I think about the forthcoming vendange, it's with trepidation.

The vats arrive one by one on the trailer. They are deposited in what is not yet my new *chai*. The cooling machine for the dry white arrives too, increasing my general feeling of nervousness in anticipation of the vendange.

Serge is working flat out. He has laid the concrete floor in the new *chai* and in the barrel *chai*. It's definitely looking more professional. Empty, it looks huge with a pink cement floor. The colour of the malo-lactic, I tell myself.

*　　*　　*

'*Quoooqiloo, mes enfants!*' Juliana announces her arrival. The dogs adore her, the cats adore her and I do too. '*Oh, la!*' as I drop a casserole I'm holding. 'Never mind, it keeps the shopkeepers in business!' she laughs. She's brought me some home made *pistou*, she says, and she's making some *anchoïade*, an anchovy paste. Do I like it? Not everybody does. In her village in Italy, she continues, Sunday morning Mass is always followed by a gossip by the fountain. It's Sunday here, and mass is over.

'We eat *anchoïade* or *pistou* on toast rubbed with garlic while catching up with the news. *Eh alors,* here's your *pistou* and '*quoi de neuf?*' what's the news?'

She's laughing. She has just been to see Pepita to find out the latest news on Roger in hospital, and recounts how Pepita was telling her how she has been to the hospital too, for a problem with her ears. 'She said the doctor sent her to the photocopier!' laughs Juliana.

I look at Juliana, nonplussed.

She is laughing and can hardly speak and when she does I can't understand her. '*Attends! Attends!*' she splutters, holding her sides. '*Oh, la!*' and she bangs her hand down on the table. 'She said they photocopied her and now they can see what's wrong!' Juliana laughs uncontrollably, holding her stomach and slapping her thigh. '*Oh la!* For her x-ray!' I'm now laughing too.

Dear Patronne, How are the grapes coming along? Am in training, so am fit. Have shaved head in anticipation of grape skin head falling experiences. Kim's on for some too. Love, your ouvrier favori.

    PS Here's a poem.

> *Waste of Time*
> *I had a shave the other day*
> *And it bloomin' well grew back.*
> *That was a bloody waste of time.*

Edge

Gilles is going to hospital for another operation, this time at the base of his spine. He'll have to stay in for two weeks, he says, and he mustn't do anything strenuous afterwards.

'Oh, la, la, he should never have had any operations in the first place!' says Annie down at La Ferriere. Jean is filling in a form and stamping documents for me. 'I told him years ago. He should have gone to see my osteopath.' She continues 'It's a complete joke! They shouldn't allow another operation.'

Gilles doesn't look as though he saw the joke. He can't turn towards me as I approach his hospital bed. He smiles a weak smile. '*Salut*,' he says, softly. Pamela is next to him. She looks at me with her wide eyes and says nothing.

'How was he?' Michel is at my house, with two baskets filled with tomatoes, melons and green peppers, all freshly picked that morning.

'Not good', I say. In fact, he'd looked shrunken and vulnerable and I'd been shocked to see him so changed.

'Oh, you don't know Gilles,' says Michel. 'He'll bounce back, you'll see!'

Gilles's vendange will be picked by machine. The quantities are too great to pick by hand this year. However, beforehand, we'll go through the vines and hand pick off anything that's not good. That way, the machine gathers in only the very best. And Luc de Conti's machine will pick them. His cousin and the driver of the machine, Francis de Conti, tells me it's a state of the art machine which will gently massage the grapes off the vines rather than shake them.

He explains that I need tractors, pipes, pumps and two hydraulically lifting trailers. The trailer that Gilles has sold me, along with all the other equipment on the vineyard, is not a hydraulic trailer. If I want the grapes to get to the vat in a clean state, says Francis, I must have them.

My mind is racing again. Where can I get one, let alone two? How much do they cost? The vendange is almost here and there's no time to lose. I can't borrow one as everyone will be using them at the same time.

'*Bonjour Patricia.*' One second-hand hydraulically lifting trailer on credit from Monsieur Roy.

Gilles is back. He's walking up the road towards me. '*Salut!*' he says, smiling at my surprise to see him. 'I've got to walk each day'.

' . . . and stay on the road!' I finish his sentence.

He's feeling much better and *Milledieu*, he just wants to get back to normal. And no more tractor, I say.

'*Oui*', then '*au moins, pas pour le moment.*'

'*Il faut les ramasser,*' it's time to pick them. We're not even half way through September and Jean Marc and I are tasting sauvignon grapes in the vines.

They have aroma, they have good acids, they have good sugar and it's time. I have only one hydraulic trailer and only one vat that can be modified to do an overnight maceration of the grapes.

'It's worth doing', suggests Jean Marc, 'as the grapes have *matière*. Ring up de Conti and see if he can come over this morning.'

This morning? So quickly? Surges of adrenaline and fear in equal measures spread through my body.

I ring him. He's coming in an hour. I rush to Ste Foy la Grande to buy a very large bottle of carbon dioxide to neutralise and protect the vendange as it comes in. I rush to the telephone to ring Charles, who lives half an hour away, to see whether he can come and drive the tractor. I rush to the vat to make sure the system for maceration I've set up is properly in place. Then I rush to the window to watch for the arrival of the vendanging machine.

Charles arrives behind it. It is truly enormous, higher than a double decker bus and as long, with huge wheels that will straddle the vines. Enormous massaging handles that look like teeth are positioned on either side of the long but narrow aperture in its middle, it is a colossus, standing in the road and dominating the space and us. A deafening whirring noise emits from its very core.

From a great height, the driver descends to shake my hand. Which vines, he wants to know.

'Bloody hell fire!' says Charles, who has got out of his car and is gazing up at the machine, dwarfed by its immense bulk and height.

They set off into the vines. Watching them progress up the nearest row of vines towards the first *parcelle* to be picked, I wonder how this huge monster will possibly be able to turn at the end of it. The large blue and silver giant lumbers up the row straddling the vines; not picking, but intent on its journey. I look at the tractor with Charles on it, leading the way to the designated *parcelle*. It looks minute, lost in the greater canvas. Indeed, the vines look minute too. The machine turns effortlessly at the end of the row.

If it was deafeningly noisy before it started gathering in grapes, it now pitches into another realm of sound – a high, screeching noise, accompanied by what sounds like a gale force wind. It moves along the row, gathering grapes deep into its belly.

It moves slowly and deliberately. Then turns at the bottom of the row and proceeds up the next one with no hesitation, continuous movement and precision in equal measures. I look at the small rose bush that had been planted earlier in the year at the end of the row and see it is still standing, untouched by the monstrous wheels. I look at the vines and see only the cadavers of bunches of grapes. They are completely denuded of fruit, but the stalks are still intact on the vines.

Twenty minutes later and Charles arrives with the first load in the trailer; a huge delivery of translucent, pale grapes. We haven't time to be overawed by the quantity, as the pipe needs to be attached to the tap at the bottom end of the trailer to pump away the juice that has accumulated in its base in the last twenty minutes.

And the carbon dioxide has to be sprayed into the vat before any juice or grapes are sent into it. I climb the stepladder and spray it from a gun on the end of a thin black pipe, which in turn is attached to the bottle. A spray of snow fills the vat. I only need a few seconds worth of it, Jean Marc has told me. But I can't switch the gun off as the handle freezes and sticks to my hand. Snow-like carbon dioxide overflows out of the top of the vat. I climb down the stepladder with the snow, the gun and my hand frozen on to it and switch off the supply from the bottle, whilst Charles wrestles to separate my hand from the gun.

We tip the grapes into the crusher from the trailer. The noise in the *chai* is now deafening. The tractor is revving loudly to power-lift the trailer and tip the grapes into the crusher. The vendange machine in the distance continues to whirr. The crusher vibrates and grinds as load after load of grapes thud in and hit the base of the vat.

Gilles, Juliana and Yves are standing well back, watching as the grapes and pips and skins and pulp hurtle up the pipe from the crusher and into the vat. It takes twenty minutes to empty one full trailer of sauvignon grapes, giving twenty hectolitres of vendange. As the last grapes are sent through and one machine after the other winds down, I climb the ladder again and spray more carbon dioxide onto the load. Charles is gone. For the moment, there is silence.

Then he is back again in less than ten minutes with another load. 'We need two tractors and trailers!' he shouts. 'He's waiting with another load!'

I add some sulphur dioxide to stop the natural yeasts on the skins starting the fermentation process. And I worry about oxidation, even though the carbon dioxide should neutralise and seal the must and skins. Even though I've already added enough of it to satisfy a whole vat's worth.

I think back to my usual white wine process and can't imagine how I managed to make anything approaching a reasonable white without doing this. Then I remember that I didn't. I was completely unaware of such essentials as keeping the juice cool, preserving the fruit, acidity and freshness and preventing oxidation. At this moment, they are occupying every brain cell.

I have now mastered the carbon dioxide gun and am using it in short bursts. I spray the snow on top of the last load again, to make sure there will be no hint of oxidation. The quantities of grapes coming in are awesome and I frantically run through in my head everything I've read and everything that Jean Marc and Bruno have told me about the maceration process and what I should be doing.

Is there anything I haven't done that I ought to have?

I've added the sulphur dioxide to stop the fermentation, knowing it will be dissipated when I run off the juice tomorrow and hoping the grapes and must won't heat up and start fermenting beforehand. I am fervently hoping it won't because Jean Marc has told me what hell it will be if it does start. Indeed, the reason we're picking this morning is because we want to gather in the harvest with its freshness and aromas intact, which requires the grapes to be cool. If the fermentation starts on its skins, the resulting juice will be too tannic and bitter, with vegetal extracts and a deeper colour.

I've added the pectolactic acids to extract fruit and aid the maceration process. And I've added more than enough carbon dioxide.

Another load hits the crusher and is sent up into the vat. 'Only one more to go!' shouts Charles as he drives off. Just as well, as there is only just room for one more. With the vat full to the very top and covered with a last burst of carbon dioxide, I position the small lid over the opening.

Charles and I gape at it. It has taken less than three hours to pick what would normally take us two days.

The whirring of the vendange machine becomes louder as it approaches the *chai*. The driver comes to shake hands and say goodbye. *'Très jolie vendange'*, he says, *'très propre,'* and I'm full of pride that the harvest was clean and merited such a comment.

We taste the juice. It is utterly delicious, full of fruit and good acids and aromas and it's my first sauvignon pick from my new vineyard and I want to shout with joy that we've succeeded at least in gathering it in.

Charles cleans the crusher while I clean and clean the underground vat in preparation for tomorrow's free juice run. Shell-shocked and disjointed, I reflect on the events of the day. No sooner had we started the pick than we finished. This morning I got out of bed without any idea of picking. Less than six months ago I wasn't expecting to have any white grapes at all to pick. Now I have a one-hundred-and-fifty-hectolitre vat full of grapes and pips and skins and pulp. And not ugniblanc grapes, but sauvignon. Furthermore, it's only fifty percent of my white wine harvest, the rest being semillon and muscadelle. Which is still sitting out there.

It's morning, five o'clock in the morning and I'm about to start running off the white juice from the skins, having attached the pipe from the full vat to send into the underground *cuve*. I spray carbon dioxide snow deep into the bowels of the *cuve*, then open the tap to run off the juice into it.

The *cuve* couldn't be cleaner and the juice couldn't be clearer. It is astonishing. It hasn't started its fermentation and the fruit flavours and aromas are there. It's clean and clear and fresh and the new piping system in the vat for draining the juice has worked effectively. I watch the juice run into the *cuve*, while holding the pipe firmly onto the floor to ensure no air is sent into it, occasionally spraying it with snow because I'm now obsessed with oxidation and obsessed about making a good dry white. I taste it.

It's cool and fresh and I know, this time, I've got it right.

Two hours later and the juice from the tap is a mere trickle. Charles is just arrived with Michel, to help transfer the pips and skins to the press. We open the door to find the *vendange*, not as dry as the red under similar circumstances, but solid all the same. And we set to work digging out skins and heaving them into the press. Pipes and pumps are ready, another vat is clean and waiting and we start the press.

Armed with my snow machine from which I am now inseparable, I start moving the free juice out of the underground *cuve*. More carbon dioxide is sprayed into the clean vat, then all the juice is transferred into it from the underground vat, finishing with a blast of snow.

Monsieur Roy's cooling machine sets to work. The juice was at fifteen degrees Celsius in the underground *cuve* and I am going to reduce it to five. It courses through the machine and back into the vat, cooling without aeration. As always, pipes and a pump are hard at work. I stand back and watch it in delight and relief. I can already see condensation forming on the side of the vat as the juice begins to cool.

There are pipes everywhere: going into the cooling machine, coming out of it, going into the vat and back into the cooling machine. Going from the tray of the press into a second pump and from there into the vat. But in spite of what appears to be chaos, we are under control.

Once the juice is cool enough, it will sit for up to a week with all its aromas and fruit flavours intact. And once we've finished pressing the grapes, the rest of the juice will be sent into another vat. And we'll do the same with that. It only remains to wash out the underground *cuve*. And the vat that held the macerating *vendange*. And empty the press. And run the skins and pips in wheelbarrows to the road. And wash all the pipes and pumps and buckets . . .

'Hey, boss! You didn't tell me you were starting so soon!' Edge writes to say he'll be arriving with his new turbo-charged babe. *'And you didn't tell me we were entering into hi-tech winemaking either! Don't think I'll come. Nah, only kidding. See you at the weekend. And here's one of my poems.*

> P
> *I've an idea that his id*
> *Was a little odd*
> *Maybe due to*
> *An alter-ego*
> *He used to know*
> *(Quite long ago)*
> *Or a super ego*
> *Ergo Sum*
> *Quid pro quo*
> *Toe Tacle Tum*
> *You are what you are*
> *You walk when you run (afterwards)*
> *(And before)*
> *(And you write stuff in brackets*
> *When you're not really sure.)*
> *Edge*
> *From your ouvrier favori*

'Comment c'était?' Bruno and Claudie are here. They've come to taste.

'C'est autre chose!' pronounces Bruno, as we taste the juice. The

wine is still cold and still settling. 'If I were you, I'd filter the *leas* too' he says.

Jean Marc had already mentioned that when he called yesterday. '*C'est autre chose, n'est-ce pas?*' he had said with approval. He suggested leaving it a few days longer before finishing the cold settling, and injecting the cultured yeast only when the juice naturally reaches 15 degrees.

'But watch out it doesn't start fermenting before you've taken it off its *gros leas*,' he cautions 'And make sure you filter the *leas* before they start fermenting too. And be sure to control the temperature while it's fermenting as it mustn't exceed nineteen degrees; in fact, if you can keep it at seventeen, all the better.'

'Cool, boss!' Edge and his turbo-charged babe are here and inspecting the new *chai* and equipment whilst tasting the cold sauvignon. He has no hair this year, apart from a small tuft at the top of his forehead. His turbo-charged babe does. Hers is blonde and shoulder length, framing a small, pretty face. Edge seems to have grown taller still and is wearing multi-coloured long trousers, no shoes and his usual tee shirt with *Green Power* sprayed on the front.

'I'm in training, boss,' he says, 'and I'm a meat eater for the moment. But maybe I'll switch back to veggies. Or maybe not. We'll just play it by ear, shall we?' We laugh. 'So where's the wood then? Come on, Kim, let's get to it.'

The filtering man is here to filter the *leas* and we've added the cultured yeast to the sauvignon; a special yeast for low temperatures. We are also about to pick the semillon and muscadelle grapes for the rest of the dry white. And we're going to put some of the sauvignon into barrels.

'How come you decided to go for it all in one hit, Boss?' asks Edge, watching the filtering man set up his machine and gather together most of our pipes. The filtering man said he could only come this morning, or not until the end of the week.

Francis de Conti has rung to confirm he could change things around to accommodate a morning pick. The sauvignon vat is fermenting, and if we want a barrelled white as well as a classic white, now is the moment to fill barrels. The characteristic whirring sound announces the arrival of the vendanging machine.

'Wow!' says Edge, looking at the huge machine sitting in the road. 'The dog's bollocks!'

Eight hours later and two vats are full to the brim with semillon and muscadelle grapes, the *leas* are filtered and cooled and ten barrels of fermenting sauvignon juice are sitting in the barrel *chai*. The *chai* is clean with wonderful aromas of fresh fruit, gooseberries and elderflowers as we close the door on a hard but rewarding day's work.

Edge's appetite has not decreased. 'Yeah, bit of a carnivore at the moment,' he explains. 'I'm doing a run for the blind when I get back. And I'm in pre-training for the marathon. And I'm rowing again.' He has moved on to the cheeses and is demolishing a ripe Camembert. 'The dog's bollocks, this,' he says, scooping another spoonful onto his plate.

'Did I tell you I applied to *Blind Date*?' He laughs as he tells me how he queued up with countless other hopefuls outside LWT and was eventually shortlisted. 'Yeah, and the *Daily Mirror* interviewed me and one of my mates while we were waiting. It was pretty cool.' Attacking the apple tart, he continues 'I thought I gave a good interview, actually. I made it very *Daily Mirror* to make sure they put me in. They asked what sort of girl I would like if I were chosen and what sort of questions I might ask from behind the screen, all that sort of stuff. And they took pictures of me too. So next morning I rushed out to buy loads of copies of the paper.' He explodes into laughter again. 'When I read it, I sounded like a real nerd. And it was too late to buy the whole print run before my mates bought it. I'd already told them about it in the pub the night before.'

A day from hell. At 5 am I attach the pipe to run off the semillon juice into the underground vat. The aromas from the fermenting sauvignon vat are more concentrated than last night. I climb the ladder to inspect the juice and insert the thermometer. The noise of fermenting yeast is intense inside the vat and I know before I take a reading that it has increased in temperature. I'm not wrong. It's at 18.5 degrees and has to be cooled down to 16 and right now.

Out with the pipes and the cooling machine. Two hours later it begins to descend in temperature, but slowly. The barrels of sauvignon are overflowing slightly too and the cooling machine will be needed for the semillon and muscadelle juice as well as for the fermenting sauvignon. Oh God.

Charles and Edge are in front of the semillon muscadelle vat and opening the door. 'No! Close it!' I shout, too late. They both leap out of the way as the door bursts open and tons of grape skins and pips burst onto the floor, along with the residue of juice left in the vat. We stare at the translucent yellow mountain in front of the vat. It grows before our eyes as more and more *vendange* oozes out and extends further into the centre of the *chai*.

In a race against time and oxidation we shovel the grapes into the black dustbins, with liberal shots of carbon dioxide. Edge throws the bin-loads into the press while Kim assembles the pumping equipment and Charles and I keep shovelling. Thank God for Serge's floor, Charles's staying power and Edge's muscles.

Thank God, too, for the electrician who arrived within minutes of my emergency phone call after the third power cut in succession, due to an overload on the electrical system. For a while we had no cooling machine, no pumps and no press. I stood in a silent *chai* panicking at the prospect of totally oxidised juice, with fruit flavours gone and all my hopes of a white wine disappearing.

'What about a pizza then, Boss?' suggests Edge after the press. We head down to the pizzeria at Gardonne as we are, grape skins, wellies and thoroughly relieved to have pressed the grapes. And we haven't even started the reds yet, or the rosé or the sweet, I tell myself as we drive down the hill.

'Décidément! Il y a un changement ici,' says Geoffroy, looking around the *chai* admiringly. Every vat is full and new barrels have just been delivered for the fourth noble rot pick. Edge and Kim are about to leave, having managed two hand picks before the machine picks on the reds, as well as the actual machine picks and three noble rot days. We have also done plenty of pump overs, temperature and density readings and scissor sharpening, to say nothing of *hotte* carrying and loading the presses, emptying and transporting the skins to the road and cutting wood.

'Really sorry to be going, Boss,' says Edge. 'But I'll be back!'

I'm really sorry they're going too. His loyalty, companionship and energy are boundless.

The amounts of wine I now have in my *chai* are industrial quantities compared with what I had before, and the equipment is also an improvement. The cooling machine for the whites doubled up as a heating machine for the reds. The malos are finished and the wine is in barrels, eighty-five of them, some of them new. And if I haven't quite perfected the art of filling barrels without wasting a drop, I'm much better able to calculate when I should stop the pump, and have become adept at topping up with short bursts of the pumps without sending wine shooting up into the air.

I am on my own in the *chai* after a day spent filling barrels. I am wet and cold, but have a great sense of achievement as I look at the new barrel *chai*. The barrels of red wine are now full, sitting in rows two high, with the dry white barrels alongside and the Saussignac fermenting gently in oak at the far end.

It is extraordinary to think that we managed to pick all the whites, then the reds and the sweet and get to this stage without too many disasters. It is also extraordinary that I have sixteen hectares of vines.

Geoffroy is right. Things are very different.

The only wines left in vats are the rosé and the classic dry white. Fermenting them was fraught with periods of panic. The cooling machine had to work flat out on the sauvignon vat, then the semillon, plus the muscadelle and also the rosé. Working day and night was tiring and taxing, but we were kept going by the adrenaline of the moment, and probably the fear of impending disaster. Every operation was immediate with no time to spare – particularly the mountain of grapes on the floor of the *chai*. And yet we managed it all. Just like every other winemaker.

The whites are delicious. Each time I taste them, I marvel at the difference between what I have just made and what went before. I marvel, too, at the difference between the barrelled white compared with the one in the vat. They were picked at the same time, from the same *parcelles* and treated in the same way, but putting some of it into oak has made two very different wines. The white in the vat is fresh, lively and fruity, waking up the taste buds instantly with lemons and flowers. The barrelled white is developing flavours of grapefruit, melon, vanilla and rich, toasted fruits.

And the rosé has an overpowering aroma of raspberries and strawberries. Equally lively, it hits the tip of the tongue with freshness and vitality.

Is every other winemaker feeling like me at this moment, I

wonder? Happy to have the grapes safely harvested and inside where there is some sense of control, even though the wine is still vulnerable? And does every other winemaker have wonderful helpers like Charles and Edge and all my pickers? And is every other winemaker feeling as utterly exhausted as I feel now?

'Coooquooo!' Juliana is beside me. *'Oh la, la, tu travailles comme une folle!'* It's 10.30 at night and she's brought some hot, freshly made tagliatelle with *trompettes de la mort* and breast of rabbit, 'to tempt you out of the *chai*! And you're soaked as usual!' she continues. 'Even the cats are fed up. I've just fed them for you; they got tired of waiting.'

Edge has left his Doc Marten boots behind. *'Ma chère Patronne, Ran the Marathon. Ace. Pity you couldn't have been there to see me. Hang onto the boots. Will collect them in spring. Guess there might be something to do then. I'll be in training. Here's a poem:*

> *Sexual Equality*
>
> *Men wear trousers*
> *Women wear dresses*
> *Men eat their dinner*
> *Women do the dishes*
> *Women have babies*
> *Men have balls*
> *Women have men*
> *Men have to watch football*
> *I make that 4 all*
> *Ie Equal*

Edge

PS Think this is the biggie with Kim. Even thinking the unthinkable, the big M.

*   *   *

The rich, golden plumage of autumn is waning. The vines, now denuded of leaves, still have a mysterious beauty. Their forms are stark against the mantle of swirling grey cloud behind them. The leaves, having nourished the harvest, are strewn over the earth, covering it with a carpet of pale gold and beige, in contrast to the leaden sky. The few grapes left on the vines hang black and wizened, some with the noble rot I'd been waiting for and given up on. I now give them up with relief, calculating that, in total, there would hardly have been a small press's worth anyway.

Madame

Vous êtes convoquée par la commission de dégustation

Vous êtes priée de bien vouloir participer à la prochaine commission de dégustation qui aura lieu le:

Jeudi 04 janvier at 10 h 30 précises

'It means just what it says' confirms Bruno. 'You'll be tasting to decide whether or not a wine can have its have its *appellation*.'

I can feel panic and fear rising, just like in the early days, but also I'm delighted to be asked. 'How do I do it?' I ask.

'Just as you do normally, of course,' Bruno replies

The tasting room at the CIVRB is in the same building as the laboratory, one floor up. It is a large rectangular room with a series of long, narrow windows looking out onto the Dordogne river and the beautiful old bridge of Bergerac. It has high ceilings and bare stone walls.

Each taster has a small screened area with a basin for spitting and a light to examine the clarity or depth of the wine. The tasting stations are arranged in a semicircle in the centre of the room with two arcs on each side of it, like a modern arena. There are thirty-three places in all, arranged in sets of three – enough for eleven 'juries'.

'*Ah, tiens! Salut, Patricia.*' Luc is here, as is Christian Roche, Bruno and Jean Marc as well as officials of the INAO, *oenologues* from the CIVRB and other *convoqués*. Every jury set has a series of fifteen or so unopened bottles. We are split into groups of three and given a number, a form to sign and another on which to make notes.

Feeling distinctly nervous, I look at my form. The different sections are: number of sample, appearance, olfactory elements and gustative elements, *harmonie*, faults, comments, and five sub divisions in the marking section, ranging from excellent to unacceptable.

I have never met the two other members of my jury before, although one of them I recognise as an *oenologue* at the CIVRB. '*Bonjour, Madame,*' he says and shakes my hand, as does my neighbour. Then there is silence as each panel starts to taste and judge.

The only sound is that of capsules being removed from bottles, wine being poured into glasses and tasters drawing in air to aerate the wine. Concentration quickly dispels my nervousness.

Holding the glass by the stem up to the light, I look for limpidity or brilliance and intensity of colour. Then cleanliness, finesse, fruit and intensity on the nose. I note what I find in short bursts, then lose concentration momentarily as some wine from the bottom of the glass drops onto the form, smudging what I've written. They'll never be able to read it. I look up to see one of the INOA officials looking at me inquisitively. Do I want something? he seems to be asking with his eyes.

I taste, drawing in air to help aerate the wine and send all the elements up towards my nose. The balance of fruit and acidity in the first sample is evident from the initial impact in my mouth, with harmonious flavours, cleanliness and a return of fruit after I spit it out.

The second bottle is already being passed to me. I panic momentarily as I haven't written anything in the *gustative* section for the first sample. I decide to mark a cross next to Excellent, then add 'persistence, fruit, *très bien équilibré*' and it's on to the next.

The bottles come in quick succession, some good, some undrinkable. Then the tasting is over. There hardly seemed time to make any notes on the samples.

I think about what I've just done. It's all very well to know that *dégustation* is a matter of individual taste and that you must believe in your ability to trust your palate and recognise certain characteristics. But I also know just how difficult it is to make one of these bottles of wine. And in spite of that, I have just scribbled unacceptable on three of them. One of them had the unmistakable smell of reduction, bad eggs, not improved even by swirling the glass to aerate out the sulphur smell. Another was green, acid and astringent and the third had the faint smell of vinegar on the nose which came out again in the mouth.

We are called to the front to comment on the wines. The INAO official asks our panel whether sample one passed or not. We all agree it did and we all agree it was excellent. 'Sample two?' he asks. '*Oui*,' replies one of my panel. The official looks at me for a response. '*Non*,' I say, and feel myself shrinking in embarrassment with burning cheeks.

'*Pourquoi pas?*'

'*Réduit*', I say in a small voice, '*et un odeur de logement.*'

'Yes, me too' says the *oenologue*. He has also refused it.

'I had it on limit,' adds the other member of our jury.

'*Refusé*,' says the official. 'Sample three?'

Even with Michel helping, it's not going to be possible to prune all the vines. Roger is not a considered as a possibility. 'You need René,' says Gilles. 'I used to use him. He's a fast and efficient pruner and you'll get your money's worth.'

René is at my door. He is in his late fifties and impressively tall. His hair sticks out at right angles from his cap and he wears glasses that are impenetrably dirty. One of the arms is held on with Sellotape. He smiles and introduces himself. He has only two teeth and he talks in short, staccato bursts, nodding his head and gesticulating wildly with his hands and arms.

'You need someone right now! We'll start with the cabernet sauvignons!' he shouts. 'And you needn't have any worries! I've pruned all my life! Two *lats*, thirty buds! You can just leave me to get on with it!'

Gilles is here too. '*Non*, she does it differently' he says patiently. 'She's pruning shorter.'

'*Quoi!*' René exclaims, 'She won't have any vendange!'

A discussion follows in which I play no part. They shout and chatter and talk about me as though I am not there and eventually reach a conclusion of sorts. I explain that I know I can't prune short in one season, as I will simply have too much leaf growth, with bunches of grapes too close together. And I know this will favour rot of the ignoble kind. All the same, I want them shorter. And I want only one *lat* as I want less vendange anyway. René shakes his head out of pity for me.

Beth arrives into the world on the 5th February, at record speed. She is small, beautiful and loved by everyone. Blonde, like Amy Louise, she sleeps and smiles and wakes and sleeps – a model child. My visit to Chantal and Damian is a short one, as pruning, racking and *ouillages* in the *chai* are all pressing.

'I'll be out in the summer, Mum,' smiles Chantal. 'Don't worry, they won't have changed that much. They'll just be a bit bigger.'

It is cold. It is very cold, minus ten. The landscape is ravishing. As I look out of the kitchen windows, the mulberry trees along one side of Eida's field are hung with white diamonds, the ground sparkling white against the dark green of the cypress trees in the cemetery. The sky is blue, a clear, vivid blue with not a breath of wind.

Eida is standing next to the hedge. As I cross the road to her, the cold hits my face and hands, crisp and bitter. Steam blows softly from her nose as she greets me and stamps her foot. She has frost on the soft down around her nose and mouth and on her mane. She chews the bread I've given her, masticating slowly and relishing every crumb, then jerks her head towards me in impatience.

I give her the rest of the bread and some stale carrots, and watch Edward progress across the field towards me, following his favourite path close to the mulberry trees, occasionally glancing backwards to check on possible intruders in his domain. He stops at Eida's feet and looks up at her, then at me. With a bound, he leaps towards the hedge and through it, scurrying across the road into the warmth of the house.

'*Bonjour, Eida, bonjour, Madame.*' It's Madame Cholet on her way to the cemetery. She pauses. 'This weather is good for the vines, you know. It kills all the germs. And the land needs to sleep to gather its strength for the spring.' She nods her head knowingly and continues on her way, hips swinging, bucket on her arm. Eida watches her and turns her head to what looks like an impossible angle, then snorts and gallops off.

I collect some of Edge's carefully stacked wood, the bark scratching my fingers as I throw it into the wheelbarrow. I look at the vines, stark against the whiteness that stretches as far as the eye can see. In the far distance, I see Rene, working speedily. His head bobs up and down as he works, talking to himself and moving on steadily from vine to vine. Over his shoulder hang the long-handled secateurs, balancing precariously and moving in rhythm with his head.

My wheelbarrow full, I rush back to the house, hands and feet frozen in spite of Madame Cholet's toe-warmers. I fill up the kitchen stove with wood and leave the door of it open to speed up the burning flame, which the new wood has dampened with its frost. I warm my hands in front of it. They throb painfully as feeling returns to them.

I seem to spend an inordinate amount of time trekking to the log pile behind the winemaking *chai* with the wheelbarrow. It takes a ridiculous amount of time to fill it, return to the kitchen door and empty it armful by armful into the log basket. I've given up taking the log basket to the wheelbarrow, as it quickly becomes either too heavy or the bottom falls out of it. And, with the door open, by the time I've finished, the kitchen has lost much of its heat. Then it's back to the woodpile to repeat the whole exercise again to get the longer logs needed for the *salon* stove. At some point, I must get some other form of heating instead of two stoves for the whole house.

In the meantime, it's back to the cold *chai* to rack the red barrels. I empty ten barrels' worth into a vat to get a reasonably homogenous mix, rush to the lab with a sample, then rush back to empty the rest. Work is interspersed with trips into the house to add more logs to the fires. Inside, the dogs lie in their favourite spot, spread-eagled in front of the *salon* fire. Sam opens one eye, wags his tail half-heartedly and closes his eye again. Luke, sighing, moves his leg slightly to allow me closer to the fire to throw another log on. The cats are spread out between the dogs, forming a mass of limbs; heads and tails of different sizes and species in a strange symbiotic whole.

In the *chais*, the cold swells my hands as I empty and rinse out barrel after barrel and my nose drips, a steady flow of pure water. The sleeves of my jumper are cold and wet and my feet have turned to ice inside my wellingtons. Madame Cholet's toe-warmers have long ago given up the unequal struggle against water and cold.

*   *   *

As I connect the freezing, metal ends of the pipes to the pump and start the process of filling the barrels again, night falls – and with it the temperature, which hasn't risen above minus five all day. I kneel next to the pump and watch each barrel from a distance, hand on the button ready to turn off when the barrel is full, and ear finely tuned to the sound in the barrel that will tell me when it's nearly there. My sleeve ends have turned icy cold and my feet have lost all feeling.

'*Patricia, tu es folle. Tu va attraper froid!*' Juliana is next to me, wrapped in a mohair coat, hat and a large black shawl. She gives me a kiss. Surely I don't need to be in here at this hour, she says, looking at me with sympathy and worry. Why can't it wait until morning?

It can't, I tell her. I can't leave the barrels empty and, anyway, it's nearly finished. Which it isn't.

And have I eaten? No, but I will soon, when I'm done.

Driven now by perverse and dogged determination, and spurred on by the thought of a warm *salon* and kitchen, I finish the job, and wash out the vats, pipes and pump. My clothes are wet and I am numb and purple with cold as I close the door of the *chai* and head to the house. Opening the door of the salon, the fire has only embers, with dogs and cats still sleeping in front of it. In the kitchen the stove has burnt out completely, as I forgot to close its door when I last filled it.

'*On y va!*' It's 8.30 in the morning and Odile and Peggy are with me to pull the wood.

'*Que ça fait du bien!*' says Odile as we pull and lay, pull and lay. And it's true. In spite of the cold, it's good to move and stretch one's body. Within half an hour, hats and scarves are removed as we warm up with the effort.

It is another beautiful, crisp and frosty day with a blue sky and no movement in the trees. Or, come to that, in Gageac. Everyone seems to have gone to ground.

The silence is as beautiful as the landscape, the only sounds the creak of the wires as the wood is wrenched from them, or the occasional desultory conversation of Peggy and Odile who are working at the same speed on separate rows, but further down than me. I watch them as I work towards them and feel humbled by their loyalty and friendship.

'*Quoquilooo!*' Juliana is striding towards us, dressed for work in diamond earrings and her rings and waving some secateurs. '*J'arrive!*' and she sets to on the next row. '*Qu'il fait chaud!*' and she laughs her raucous laugh. '*Tu ne sais pas la dernière!*' and she recounts the story of the visit of her friends from the Midi, who have left that very morning.

'But I thought they were staying another week?' I say.

'*Oh, la!* The Gendarmes will have taken me away by then. She's eating me out of house and home. And she and her son drink five litres of wine a day! And then they start arguing with us! The other day, when they thought I was going to throw them out, they claimed to have lost the key to their front door!' She laughs. '*Non, il faut être sérieuse, quand même!*'

Chère Patronne

Comment ça va? That's enough of the French stuff. Thinking of you. Thanks for the money for the blind. Is the tasting room finished yet? And the wall? Hope all the fermentation is done and you're having a rest. Guess you need one. I do too. Went to the pub last night and drank a bit. Here's a poem I wrote when I got back.

The Correct Pronunciation of:

Shuddy bowl
(Sugar bowl)
Shooody-bough
Schoudi bou
(Ronnie spoons)

Shoooowdiboo
Shoe-de-bu
Shoddy bou(h) (silent 'h')

Edge
Your ouvrier favori

'*Salut.*' It's Michel who's come to help Rene tighten the wires before tying up the *lats*. René is rushing up and down the vines, head bobbing, checking wires and stakes.

Michel smiles. '*Tu es libre ce soir?*' he asks. '*On peut grignoter ensemble avec Gilles et Pamela,*' and he's gone, pulling wires and uprooting stakes.

The work on the tasting room is progressing, but slowly. The breeze blocks are up and I need no longer worry whether I chose the right place for it, as it's now a fait accompli. It sits opposite the *chai* doors and will eventually be connected to the *chai* by an arch. One small tree had to come down to build it, the tree beside which Charles and James constructed the shower in the early days. The shower was dismantled as soon as the bath was operational in the house. It now seems the distant past. Edge cut the tree down, agreeing to do so only because it was half dead anyway.

In fact, there will not only be a tasting room, *salle de dégustation*, but also a *dépôt de stockage*, to alleviate the constant problem of where to store cartons of wine. The building will also house an office. At the last bottling, Monsieur Capponi and his workers crammed pallet after pallet into the atelier before depositing twelve homeless pallets at Geoffroy's. 'You really need more space,' Monsieur Capponi said, then added: 'You're not the only one. Everyone has the same problem. There's never enough space. Just sell more wine!'

'*Milledieu!*' shouts Gilles, grinning and laughing. We are having dinner at Michel and Monique's again. Pamela is laughing too.

Michel's *camionette* is a write off, he's told me, a wild boar has rammed into it again.

'How?' I ask.

'Well, there are a lot of them about at the moment,' says Michel, reddening in the face, but smiling and looking sheepishly at Gilles, then at Monique.

'Oouii!' says Gilles, touching his nose and giving a knowing look.

'But were you standing nearby? Weren't you frightened?' I say.

'Oui!' he says. It was suddenly there, right in front of him, blinking his small red eyes and champing its jaws, while pawing the ground, a hideous looking creature. He leapt out of the way at the last second and the wild boar careered into the van.

'*Tu est fou!*' says Monique in disgust. 'You could have been killed! And what about the repairs?'

Everyone is laughing. I look at Michel, who is smiling and looking from one to the other of us and try to imagine him in a town environment and can't. He is country through and through – his solitude, his habits, and his unconscious love of nature and the seasons are obvious to everyone except, perhaps, him. And his generosity and sincerity go hand in hand with them.

The white and the rosé are being bottled. The labels, capsules, bottles and cartons are here, along with the corks. Monsieur Capponi and his lorries are also here, with the usual noise, activity and chaos. By the end of the day, it's done. I look at the bottles and feel intensely proud. The rosé is a spectacular colour, glistening and brilliant, and the white looks wonderful, diamond bright.

My pride is replaced by panic as I realise how many cartons are stacked in front of me. Where can I put them all? Why didn't I get the *dépôt de stockage* finished? I've now got to sell them quickly, as Monsieur Capponi had suggested. And to whom? My only contacts are Justerini & Brooks, Corney & Barrow and John Davy.

I must send them samples, I must find more contacts. I must find more time.

Spring has crept up on us. Suddenly the ground is erupting with greenery. The trees are gently unfolding their leaves, fresh and beautiful, as the sun showers warmth on the land, awakening the vines and bringing new life and colour. Juliana is with me most days in the vines, tying up *lats*, along with Michel who appears and disappears as usual, and Pia who arrives from time to time with Balthazar, her black Labrador, having walked over from Saussignac. Roger is *hors de combat* again, drying out in hospital.

My days come and go in the vines, and I fall into bed at the end of each one exhausted with a healthy tiredness, and dreaming of vines, buds, small escargots and the task ahead. After tying up lats, it's mowing, cutting off shoots, lifting the wires, decrocheting, trimming the vines, weeding, spraying . . . Plus finding a market for all the wine I've already made.

Chapter 12

'CE N'EST PAS VRAI!' MADAME QUEYROU IS HERE. 'YOU CAN'T HAVE BROKEN your leg!' She bends down to kiss me. 'I came as soon as I heard.'

I have indeed broken my leg. Not scaling ladders in the *chai*, not heaving barrels and climbing over them, not driving tractors or rushing through the vines lifting wires and attaching them to nails, but visiting the accountant in Bordeaux and falling down five shallow steps in his hall.

In fact, it was not even a fall, but a leap into the void. Turning towards him to say goodbye, I opened a glass-fronted door that led into the hall and next second had landed at the bottom of some stairs I didn't know were there onto my right leg. With such force that I broke it, in three places.

'*Oh, la la! Patricia! Qu'est-ce que t'as fait?*' Juliana is kissing me and looking at my leg. '*Ce n'est pas possible!*' It is, however, entirely possible. I am still in shock, and pain.

At the hospital, the surgeon, Olivier Bigard, who is also a friend, examined my leg and then looked sternly at me. 'I ought to put pins in this. If you listen carefully and do exactly as I say we'll try something else. But you've got to stay in bed for three weeks with your leg up and have total rest. So either you do it here at the hospital or at home. Are we clear on that?'

We are.

It is hot and I am in bed at home with a purple plaster cast on my leg, stretching from my ankle to just above my knee. Cushions support my leg and I have a large black swelling around my knee, where one of the fractures is. I know from Olivier that my knee and ankle took the force of the break. The other break is just above ankle level. Each time I lower my leg, it instantly swells up and throbs, tightening the plaster cast and turning my flesh a strange, dark purple, matching the cast. A nurse will come every other day to inject me with blood-thinning agents. How could I have done such a thing?

'Right,' says Pat Chaffurin. 'Juliana and I will bring you break-fast each morning. You're not to get out of bed!'

'And I'll come over for the suppers!' says Odile 'And bring a *blanc cassis* of course,' she adds.

'Salut.' Gilles and Michel are in my bedroom, dressed only in shorts and trainers. It is hot. Very hot. Gilles, arms folded on his chest with Michel standing next to him, says 'We've been talking, Patricia, Michel and me. And we've decided that, between us, we'll do the rest of the work in the vines. You've finished the *attache*, lifting the wires and the first *épamprage*, so we'll do the spraying and *éppointée* from now on.'

Tears roll down my cheeks.

'Milledieu!' says Gilles. 'What did we say?'

Ma chère Patronne
Heard the news. What a bummer. Just as well you're mega fit. You can hack it, but it's tough. I know cos I did it once. Think Karma, read books, get better. Will have phoned by the time you get this. Hear it's really hot out there. You'll be using that fan we bought last September. Here's a poem:

> *Blown Away By a Fan*
> *One day*
> *A hot day*
> *I got some imaginary fan mail.*
> *But unfortunately*
> *It got blown away by a fan.*
> *I was really blown away.*
> *Well, not really.*
> *Imaginarily.*

Edgini Publications
(2nd edition as I've sent it to you once. Just checking . . .)
Love
Your ouvrier favori

'*Salut, ma cocotte! C'est moi!*' Odile is climbing the stairs.

It is impossibly hot and I want to get out of bed to have a shower.

'I've got an idea', says Odile. I sit on the side of the bed and lower my leg. It instantly swells to an enormous size and changes colour. '*Zut!*' says Odile, 'let's get you to the bathroom quickly.'

Sitting on Amy's small plastic stool in the bathtub my leg balanced precariously on the edge of the bath and the cast covered in a large black, plastic bin bag to stop it getting wet, Odile showers me. She is laughing. '*Oh la*, what do you look like!'

I'm laughing too, and crying.

'*Eh voilà!* Now for a *blanc cassis!*'

Justerini & Brooks
Dear Patricia
Sorry to hear about your broken leg. Just to cheer you up, here's our small order for 200 cases of your red and the same for the dry white. What was the price again?
Love Hew

*   *   *

'Grandma!' Amy is kissing me and looking at my plaster cast in deep admiration. The gorgeous Beth is sitting on the bed next to me. 'We've come to look after Grandma,' says Chantal to the girls, 'haven't we?' Becky, Edge's sister, is here too.

Amy has a white marker pen and is busy drawing the first of many crowned and berobed princesses on my plaster.

Geoffroy comes too, visiting each day. As do Nick and Eve, Bruno and Claudie, Pia and countless others.

'Quooqiloo, mes enfants! Oh la, la, elles sont belles!' says Juliana, kissing the children, then Chantal, then Becky. Juliana has brought with her stuffed moules *'pour te tenter à midi!,* to tempt you.' She smiles, looking at me, then back at the girls. The girls look at the moules in disgust and Juliana laughs and kisses them again. 'Come and see my dog Chloe,' she says to them, then looks back at me, and adds 'Not you!' collapsing into laughter again.

'J'en ai!' says Gilles. He has some crutches. 'There's no point in you hiring them. I've got my own. I've even got two sets!'

I'm allowed to get up, but mustn't put any weight on my right leg.

'Milledieu, there's not much I haven't broken' says Gilles. 'My leg, my arm, my foot! *Oui*, with my foot it was hell. You know, inside your foot it's *le vrai* spaghetti, masses of bones and cartilage. So make sure you don't put any weight on it.'

Ma Patronne
Pretty soon you'll have email, which will be ace. Hope the leg's mending. Are Becks and Chantal tanned and chilled out? How are the grapes, the little buggers? How's the noble stuff progressing? Here's a poem:

> *The Trees*
> *Does anybody wonder what they think*

When you light a campfire next to them?
Or light a cigarette?
I think they might tremble a bit.
Match
Ignition
Fear
Struck
Into the heart of the tree
Paralysed
Rooted to the ground.
. . . And I wonder how they feel
When we chop them down,
And make them into paper.
Edge
(Flat, I suppose)
Love
Your ouvrier favori
PS D'you think there's a vocation for me out there as a poet? If not, I'll settle for
a vacation.

'Mum, you're to sit there and not move,' says Chantal.

I am at last out of bed and dressed, sitting downstairs with my leg up. I feel a sense of liberation, if only partial. It's early evening and the french windows in the *salon* are wide open, looking out onto the courtyard and the wisteria. Although it has long since finished flowering, it's a mass of green fronds.

The sun is still very hot and the gentle breeze, which enters the *salon* now and then, provides some welcome respite. A lorry is arriving soon to take away ten pallets of wine for my bond-holders; their first consignment of ten cases each.

Charles has been over often to help. Not only has he calculated how much and which type of wine will be sent out, he has also marked each carton with a number and stacked them on pallets. They are sitting in the now completed *depot de stockage*, waiting to leave.

The lorry was due at 10 am, but has so far not turned up. Monsieur Capponi's fork-lift truck is also waiting outside and one of Monsieur Capponi's workers who lives locally will come over to lift the pallets on as soon as the lorry arrives.

A friend and fellow winemaker, Pierre Charlot, has just called by, delivering thirty cartons of his wine, which will be transported with some of mine on another order next week. *'Comment tu vas?'* Yes, he'd love to stay for an aperitif.

Pierre has a vineyard near Ste Foy la Grande, *appellation* Ste Foy Bordeaux. Dynamic and with endless energy, he has transformed his vineyard. His father, the former owner, was a cave co-operative member, simply gathering in the harvest and delivering it to the cave. Pierre has transformed it into a successful, working vineyard. He has a modern, well equipped *chai*, and makes red, white, and a particularly fruity clairet. Clairet was the name for a light red wine drunk fresh and cool, the original claret, he is explaining to Chantal. 'How are sales?' he asks me, 'And how's your leg?'

As he puts the glass to his lips, a man appears at the open french windows. *'Bonjour'* we all say.

'I'm English,' he replies. 'My lorry's stuck in a ditch down by the side of the château. Can someone come and help?'

Pierre leaps up and goes in search of the tractor to try and pull him out.

Someone else appears at the door, another man, holding some papers and obviously a lorry driver too. *'Vous êtes Français ?'* I ask him.

'Nup, Polish actually,' he says. 'Porter & Laker. Come for ten pallets of wine. But I've got a slight problem.' He can't get round the cross at the bottom of the road. Can someone come and guide him?

'Is it always like this here, Mum?' says Chantal, setting off. 'I thought you were only expecting one lorry?' Becky is laughing, as Chantal disappears off in the direction of the cross.

Pierre reappears. *'C'est mouvementé ici!'* he says, laughing. 'He was well and truly stuck. And just as well I brought my wine over today, because that's the order he's come to collect.'

Chantal reappears. 'Mum, do all the lorry drivers have their windscreens festooned with bras and knickers? It's no wonder he got stuck. There's no way he can see out of it! And, by the way, that cross at the bottom of the road is leaning at a weird angle.'

The vines look beautiful. Suffused with soft, morning light, they stretch away into the distance. Serene and symmetrical, they create a luxurious canopy of lush green. I walk to the merlot *parcelle* nearest the house. They have been tended with care by Gilles and Michel, who have sprayed, mowed, trimmed and weeded. The grapes are changing colour, the cabernet sauvignon in the distance hanging in long, pendulous bunches, the merlot nearby, plump and fat. I am not allowed to venture much further than the merlot *parcelle* nearest to the *chai*. But I can see that row after row has been cared for meticulously.

'Salut.' It's Gilles, who is striding towards me through the vines. He is grinning. *'Ça va?'* he says, looking around at his handiwork.

'Sam! Saam!' Sam disappears off into the distance, past the cemetery as Amy shouts and runs after him.

Madame Cazin is delivering the milk. *'Ah, oui'* she says, nodding her head. *'Il cherche la femme. Et comment allez vous?* How is your leg? I see you're now walking on crutches. What a thing to happen.'

I can put my foot down on the ground without any pain, although Olivier has told me only to put twenty per cent of my weight down for the next two weeks. I've followed all his instructions to the letter or I won't be mobile by the vendange.

'Patricia! *Quoqoo!'* shouts Juliana from her window.

*　　*　　*

'*Ca y est. Je l'ai trouvée!*' Gilles has found a house he likes, in Sigoules, about ten minutes drive away. It's right next door to where Pamela works and next to the cave co-operative.

I comment that he will be constantly seeing tractor-loads of grapes passing by on their way to the cave during the vendange.

'*Oui!*' he laughs 'Can't get away from them!' It's perfect for him, he adds, with more than enough space for his mother, if ever he can persuade her to leave her house.

He can't. She won't leave. 'I'll help you in the vines, Madame,' she says, looking directly at me and ignoring Gilles, who is standing next to us. 'And I hope you're coming to live in Gilles's house?' She turns to her son, looks at him fiercely and adds emphatically: 'And I hope you're not taking my work tools?'

'*Milledieu!*' shouts Gilles. 'See what I mean?'

The move takes place, but little by little. There's a wall to build, says Gilles, and he's not in a hurry. Better to do it well, *n'est-ce pas*? His regular trips back and forth in his *fourgonette* with stones I have given him make the road more busy than it has ever been out of vendange season.

Madame Cholet is displeased. '*Madame,*' she says, 'those stones! He's lifting stones again!'

After stone transporting trips come *armoire* and bed transporting excursions, dozens of them. Gilles is so busy he no longer has time to drop in for a coffee, so he waves each time he passes by. Prune, his black Labrador, is in the seat beside him. I can't imagine how the amount of furniture he is moving could possibly have fitted into his house.

Then it's finished. The move is completed.

Gilles is in my kitchen, having a coffee. '*Milledieu, c'était dûr!*' he says, it was hard work. He and Pamela spent the night there last

274

night and now it only remains to clean the house over here. He will bring the keys over later, he says, and come and say good-bye. 'And I hope you're coming over to see the new house? I'll make you a coffee. But I'll be coming over here each day anyway, to check on *ma mère.*'

Chere Patronne
Hope the leg's nearly better. Hear you're hopping around. Tell Becks that Chelsea are now through to round five of the FA Cup. I'll be out for the vendange. Already arranged my hols with the boss draftsman here, one advantage of being self-employed. Here's a poem:

> *My Boss*
> *Has told me to act my age,*
> *Has told me not to drink beer*
> > *out of a bottle,*
> *at lunchtime,*
> *Has told me to stop daydreaming,*
> *Has told me that I should have*
> *Put a border round a table*
> > *that I drew the other day,*
> *And a while ago he told me off*
> > *for eating biscuits.*

Edge
Love
Your ouvrier favori

Olivier, the surgeon, is holding what looks like a chain saw up in the air, revving the motor. *'On y va?'* he says, and cuts off the plaster.

Underneath, my leg is a puny, shrivelled baguette compared with the other.

'C'est normal!' he says, 'What did you expect? You haven't walked on it or used it for eight weeks. And you still need your crutch for

some time yet. And don't overdo it! And don't forget to keep me two cases of Saussignac!' he shouts as I hobble out of the surgery.

'Grandma! Your leg's all skinny!' says Amy when she sees it. She and Beth inspect in detail, then disappear into the garden with the crutches to play Grandma's Broken Leg games with them.

'Look, Beth! This is Grandma trying to walk,' she calls, and hobbles around, doing a good imitation of a wobble, then sits on the ground. They both laugh uncontrollably.

A letter has arrived from the *Guide Hachette*. I must keep its news a secret until September, but I've won another *Coup de Coeur* for the Saussignac. Immediately I phone Bruno and Claudie to tell them and to see whether they have a similar letter. They do.

'Bring the children over to meet Hannah,' they say, 'then stay for supper.'

Hannah is a six-month-old ass, a very beautiful six month old ass. The children are captivated by her, and Bruno and Claudie by them. Claudie lifts them up to Hannah and sits them on her back. They squeal with delight and fear and run around in excitement and pleasure when they meet yet more dogs, Blitz and Bises, and a cat, Petrus, who immediately beats a hasty retreat when it sees them.

It's a hot summer evening and the view from Bruno and Claudie's balcony is extraordinarily beautiful. The vines of Monbazillac stretch up and away into the distance, with a summer haze over their soft silhouettes, and a huge, red sun descending majestically towards the ridge. We gaze at it, sipping a glass of rosé, cool and fresh. The girls are quietly playing with the dogs and a general sense of well being washes over us.

'*Plein de fruit*,' says Bruno of the rosé, which is mine, and I feel proud and delighted. 'Are you prepared for the vendange in the *chai*? It'll be on us before we know it.'

'*Bonjour, Madame.* Can we come and taste your wine this afternoon? There will be five of us. I would like to present Monsieur Katsuyama, Monsieur Ochiai and myself, Monsieur Komata.' A group of Japanese have come to taste.

They taste everything, white, rosé, red and sweet. They photograph everything; the barrels, the bottles, the colour of the wine in the glass. They fetch various pieces of photographic equipment from their car, silver umbrellas and the like, and repeat the process. They enthuse about the wines and present their cards, bowing. Only Monsieur Komata speaks French and none of them speak any English. I am reminded of my first year here as I'm unable to communicate with them. They present their cards, bow and leave, thanking me.

Sales from the house have increased. As well as the locals, I get visits from television viewers who saw the documentary and are on holiday in the area. And I'm selling to a number of restaurants in the area, including l'Imparfait and the Moulin de Malfourat, where Geoffroy and I often eat.

It's time to buy bottles of carbon dioxide, yeasts and pectolictic acids in preparation for the vendange. It's still August, but Bruno's words are ringing in my ears. Time is passing and the grapes are ripening. Without my crutch, but still fragile, I walk carefully through the vines to the sauvignon grapes.

They have changed colour and are translucent, yellow and pale green pearls. The red varieties are now black; rich, heavy bunches hanging down from the vines. The heat of summer has dried a lot of the leaves around the grapes and I say a special prayer of thanks that although we didn't de-leaf this year, nature did it for us. The semillon is progressing, but there is no sign as yet of any noble rot starting. We have had a very dry hot summer and I know the *botrytis* will be late in coming, which might not be so bad, as we'll have so much else to do beforehand. On the other hand, perhaps it's not such a good idea that the *botrytis* will be late coming, I decide, as the risk of rain will be greater.

I give up on the train of thought that started with musings on sweetness and richness, an almost ideal state for picking, and led me to worry about unwanted rain and saturated, porous skins on noble rot grapes, their sugar washed away and their concentration diluted to nothing.

As I walk back from the vines, Madame Cholet's house is strangely quiet. As I pass by, I hear the church bells ringing in peals of two — a woman has died. The bells announce the death of Madame Cholet.

Gently, with three peals of two, morning and evening, they continue for the next three days. As she wanted, she carried on in her own house for the rest of her life, a year after Gilles's move. She died peacefully, aged ninety, watching television in her favourite armchair.

Her funeral is small and touching, with Juliana and Yves, Madame Queyrou and her husband, Gilles and Pamela and her grandchildren, along with family friends.

We walk up to the cemetery after the service to say a final goodbye to her with a sense of sadness. but also with a sense of a life lived as she would have wished, and a death without pain. Silence and a sense of timelessness pervade the cemetery. I look at Monsieur Cholet's plaque and his grave that she tended, now with her coffin carefully placed next it. The cypress trees, immense and imposing, stand guard next to them.

'Tu vois?' says Gilles quietly. *'Elle est restée têtue jusqu'à la fin'*, she was stubborn to the end.

'Bye Grandma!' Chantal, Becky, Amy and Beth are waving, arms out of the windows of the car as it slowly moves towards the cross at the bottom of the road, then turns right towards the château and down the hill. Juliana is beside me, waving and crying to see them go. So am I, and so are they.

The autumn landscape is spectacular; shimmering shades of gold, orange and red. The vivid yellow, red and rust leaves on the vines reflect the colours in the trees; great swathes of colour, always in contrast with the rich, dark green of the cypress trees in the cemetery. A low sun, with autumn light and mist, casts a magical glow over Eida's field as well as the rest of the landscape in front of me, ablaze with colour and soft contours.

Eida nuzzles my hand as I give her some bread. The summer is gone and with it the sun-soaked days and sultry nights. But it's done its work in the vines. The grapes look divine, with the now familiar translucent pearls of sauvignon and semillon, the rich, black velvet of merlot and the purple pendulous jewels of cabernet sauvignon.

James's visits have ceased entirely. We lead separate lives, and have slowly but surely grown apart. His life has become more and more insular and ascetic, with a recent visit to a Buddhist retreat. Mine has become more active and energetic. Our marriage is ending. Depressed and miserable, we agree that we're going to part. James is buying a house in a village in Essex called Sible Hedingham. 'Sounds like the sort of girl I ought to have married' he says. We laugh, the laughter of sadness and complicity.

'*Ma patronne! Comment ça va?*' and a hug. Edge is back, turbo-charged Kim with him. 'Sorry we couldn't get here earlier. Had a rush job on at work and had to change everything. Cool,' he continues, looking at my leg. 'Bit swollen though, boss. Okay, let's go.' He turns back and gives me another hug. 'Sorry about you and James.' Then, 'Where's the wood? Any *remontages* to do? Up the ladder, Kim!'

The whites have already been picked, and the reds. This time, the white maceration and the run-off of its juice happen perfectly. The reds, too, were gathered without incident, picked at maximum ripeness. My delight at seeing a brown pip in the middle of the cabernet sauvignon grapes, signifying their ripeness, is compensation for the throbbing pain in my leg. It swells and stiffens almost as soon as I start walking on it each morning, and taking my wellies off at the end of the day is challenging, to say the least.

I was right about *botrytis* coming late; we've managed only two noble rot picks. 'Are we on for one tomorrow, Boss?' asks Edge, as we inspect the grapes. We are. In just two days since my last inspection, morning mists and hot, afternoon sun have done their work, with whole bunches of luscious noble rot on some vines, the rest in various stages of development, all healthy and showing great promise.

Looking at the landscape and, more particularly, the vines, most of the leaves have gone, partly from our rigorous de-leafing on the first pick. The rest are falling naturally with the end of the summer.

Caught in a blaze of autumn sunlight, the vines display an array of colours. A canopy of violet and purple and pink noble rot grapes, delicate and shrivelled and ripe for the picking are highlighted against the yellow and red of the remaining few leaves, which are draped high above the grapes at the top of the vine. The vines themselves look beautiful, with the small bunches hanging like jewels, stark against the dark, twisted vine trunks.

I ring my pickers that evening. We'll start around 11 am, when the dew has disappeared. 'Thought you'd never ring,' says Pat Chaffurin 'We've been waiting!'.

'*Bien, bien!*' says Odile.

'*Oh, la. La hotte m'attend, je suppose?*' says Richard Basque.

'We'll be there on the dot!' says Charles.

Tim has just arrived too, from England.

Bruno and Claudie are planning a pick themselves. *'Oh, la la!'* laments Claudie. *'J'ai mal aux oreilles!*' ' My ears hurt!

I too, have the usual stress spots on my scalp.

I wake early the next morning to the sound of heavy rain.

'Don't worry, Boss' says Edge, looking at me as I gaze despondently at the wet landscape from the kitchen window. 'Thing to do is to keep busy. No point in getting stressed. Your head will be one large stress spot. What were you saying about these beams?' He is in the kitchen, looking up at the dark beams. 'You want them cream do you?' he asks. 'Come on, Kim.'

The rain continues day after day, as does the work in the kitchen. Plates and cups and casserole dishes and pots are all deposited outside the kitchen door in the rain, while chairs and cushions, breadbaskets and pictures are moved to the *salon*. 'Quite a few holes and cracks, boss,' says Edge, inspecting the beams. 'Think we need a bit of filling.'

I put the phone down after a conversation with Bruno. 'Have Claudie's ears fallen off yet?' Edge asks.

Bruno and Claudie also can't *vendange*. At least I have reds, whites and rosé safely in my *chai*, and two picks worth of noble rot. Their vines are uniquely for sweet wine, their luscious Monbazillac.

'And how's your head, Boss? They haven't come down your forehead yet, those *boutons*, so there's hope.'

We paint beams cream and ceilings white for an entire week as the rain continues unremittingly. Edge coughs a lot.

'Nah, it's not the paint,' he says, when I ask whether he's allergic to it. 'I've had it for a while. Can't get rid of it. Probably go and see a doctor about it when I get back. Cor! Look at her eating that bird! She just ate the head, for God's sake!' he exclaims as Lulu, the cat, devours a bird she's brought in. It's gone in record time. She leaves nothing, not even the usual hearts and livers that the other cats deposit under the table as presents.

'Did you see that, Kim? Gross!' He picks up Lulu and looks her in the eye. 'What happened to peace and love?'

It has at last stopped raining, but the grapes are sodden and swollen with water. I look at them in despair. Having waited and waited, inspecting them daily for enough pure *botrytis* and delighting in its development, it has all now been obliterated. From the initial golden grape skin colour, to brown, then the most delicate shade of pink and finally, violet with the white down shroud of perfect noble rot, there are now only sodden grapes.

Wet, dripping and swollen bunches glisten as a watery sun appears from behind the clouds and shines weakly on them. The accompanying morning mist creates yet more water.

'Don't worry, Boss,' says Edge, 'they'll dry out.' Then he adds, 'Probably?' as he puts his arm round my shoulder.

To restore anything like the concentration I had last week we need more sun and a very drying wind. And it's already late October. I look closely at a few bunches to see whether any of the grapes have split, whether there's any hope at all.

'Who's for tea, then?' says Edge. 'I wrote a poem about that y'know,' he says as we walk back towards the house and the now finished kitchen.

> *Tea Time*
> *Does anybody want a coffee?*
> *Anyone?*
> *I'm getting them in!*
> *No one?*

Last chance.

I'll just get myself one then.

Edge

'Whad'ya think?' he says as I laugh. 'Thought so' he says, smiling and making the coffee.

Juliana arrives with a basket heaving with mushrooms. *'Quooqilloo mes enfants! Quel temps, Patricia. Horrible pour toi. Mais, il faut voir les côtés positifs, c'est excellent pour les champignons!'*

She is wearing her bright yellow jacket with matching yellow wellies, large gold and pearl earrings and is carrying a long stick, which is bigger than she is. Edge asks her if it is a vital tool. *'Mais oui*, to test the land for traps, move the leaves and find the mushrooms,' she explains.

She has been up since 5 am, she tells us. She has in her basket chanterelles, small, deep ochre-coloured mushrooms. Amongst them are *pieds du mouton*, large white mushrooms *'avec un goût très fin, très, très fin!* With a very delicate taste!'

But her pride lies in having found *cèpes*. *'Pour toi, Patricia,'* she says, and places the basket on the table. The *cèpes* are enormous, two of them measuring about a foot in diameter. Their smell is unmistakable. *'Non non! J'en ai encore!'* I've got enough for me as well, she insists when we say we can't take them all. 'But clean them well, Kim. Take the maggots out first' and she laughs her raucous laugh and slaps her thigh as she gets the expected reaction from Kim. *'Non, je rigolle!'* I'm only joking! and with a kiss and a wave of her hand, she's gone, returning for a moment to ask when we might *vendange*.

I continue to be invited to taste for the *Appellation Origine Contrôllée*. I now write down comments with confidence as I look, smell and taste. My vocabulary to describe what I'm tasting is no longer stilted. Nor am I unsure about writing a cross in the unacceptable column if a wine is faulty.

'*Salut, Patricia.*' Luc is talking to a group of tasters. 'I was just saying that we should be more severe in our tasting. If the aim is to improve the quality of AOC Bergerac wines, it's the only way forward.'

I consider what Luc has said. The laws of the AOC are formulated by the growers, which are then supervised by the INAO. The AOC is supposed to be not only a guarantee of origin, but also of a certain level of quality for the region.

'*Oui*,' Luc says. 'It's obvious we have to improve the quality of our wines. We can't be complacent about it. And we have an opportunity. We have just as good a *terroir* as Bordeaux. The blind tasting we did with Bordeaux and Bergeracs the other week was a good example; a lot of the Bergeracs came out better than the Bordeaux. And it's up to us to do it. We can start by not allowing mediocre, bad wines through at this level.'

'Hey, Boss, you've got black teeth. And a black tongue.' I am back from the *dégustation* where I've tasted reds, and Edge and turbocharged Kim are eating breakfast.

'Went to have a look at your sticky stuff,' he continues. 'Dunno what we're looking for really, but they're a bit rotten again. How was the disgusting? And Pia rang to invite us all to dinner tonight. Said I thought it was cool, but would check with *la Patronne*. And don't forget to keep tomorrow free,' he adds, 'for my birthday pizza. Think I'll have carbonara to start, then a pizza, and maybe then some more carbonara, followed by a *chocolat liégeois*.'

We've had almost a week of dry weather with some fresh wind, though no real sun. There's a glimmer of hope, in that the grapes are drying out, but slowly. However, as they dry out, flocks of birds appear that I haven't noticed before. They sit on the telephone wires above the *parcelles*. At first, I assumed they were gathering for an exodus to warmer climes. But they don't leave. Are they experts in noble rot? Are they waiting for the perfect grapes too? As I look up at them, a flock leaves the wires and swarms around, suddenly low and menacing. Maybe I am adding paranoia to stress *boutons*.

'*Goddag!* Welcome!' says Pia. Pia suits her name; gentle, healthy with star-bright eyes and radiant skin, burnished by sun and wind.

Her husband, Ekan, is a tall Swede with a deep, booming voice, a wide smile and a sense of humour. He strides towards us. 'Hey hey, welcome!' They live in one of the two towers of the château of Saussignac, which they bought as a ruin and have renovated with taste.

'Patricia,' says Pia, 'I'm so sorry about the weather for your grapes. But it's good now. Maybe we can pick soon? Come, let's go downstairs and have an aperitif.'

Downstairs is reached by a large, white stone staircase leading from a high, vaulted dining room.

At one end is a roaring fire. 'Your *pieds de vignes,* Patricia,' says Ekan looking at the fire. 'They burn really well.' Pia and I had gathered all my uprooted ugniblanc and cabernet sauvignon stumps and piled load after load into her Land Rover. Gathering in and stacking one hectare's worth of one-metre long *pieds de vignes* gave us some rigorous exercise and a great sense of achievement when we finally finished. Now they are burning in the huge and magnificent chimney place, radiating hot, red heat and emitting a perfume that only vine wood gives.

Kim has not met Pia and Ekan before and is captivated by the château and by their Scandinavian hospitality. Both Pia and Ekan cook, and the food is exquisite.

'Ho, Patricia! Now we teach Kim how to drink aquavit!' says Ekan in his booming voice, laughing, and holding up his glass. Kim holds hers up too, looking from Edge to me. 'You have to sing a song now!' says Ekan, 'then drink it down in one go!' He sings a Swedish hunting song. 'Skol, Kim!' and his aquavit is gone.

She looks first at us sheepishly, then the glass, then swigs it down in one gulp too.

'Cool, Kim!' says Edge. He does the same. I demur. Pia sips hers.

'Do you smoke, Edge?' asks Pia.

'No,' he says. 'It's a summer cough that's hanging around.' Then he adds, 'You shouldn't smoke either, Pia, it's bad for your health.' He, Kim and Ekan shout 'Skol!' together and down another aquavit and Pia lights her cigarette.

'God, Boss. I'm on the wagon from now on,' says Edge the next morning.

'Why did you let me drink that stuff?' adds Kim, sipping hot water and lemon and looking jaded.

'Just as well we're not doing the sticky stuff today, then,' continues Edge. 'Think I'll write some letters; better still, some poems.'

'We're picking today,' says Claudie on the phone. We've got some good *pourriture* and we're going to gather it in.'

My *botrytis* is coming along, but is still not concentrated enough. The sun is shining with a good, blowing wind, perfect to dry out the grapes.

We head off to Bruno and Claudie's and the *Champs Elysées* to pick their noble rot. They have some delicious noble rot, much further developed than mine. As we pick, the wind runs down between the rows, up and round the bunches and down the gentle slopes of their vines, drying and concentrating the grapes. I pray that the same thing is happening over at Gageac.

We are back home after a strenuous day picking, and washed and showered. Edge is scribbling notes, sitting at the table whilst talking on the phone to Becky, his sister. Kim and I are waiting for him to finish his call before driving down to Gardonne for his birthday pizza. He is twenty-nine and has had countless phone calls since our return from Bruno and Claudie's vines, plus cards and presents that had been stashed away by Kim and me.

'Yeah, had to work, even on my birthday,' he's saying to Becky, smiling and looking at me. 'It's *la Patronne*, a regular slave driver. But we're off for some nosh now.' He hands me a piece of paper, as he chats to his sister. 'It's today's inspiration,' he says.

> Lies
> Today I looked in the mirror.
> I looked older than I'd ever done before,
> I checked on a calendar.
> It was True.
> And they say that the camera never lies;
> I saw a picture of me looking really young
>   Yesterday.

Edge

'You need to know the news, Boss,' he says in the restaurant after we have finished our pizza and are waiting for the chocolat liegeois, of which Edge is having a double helping. 'Kim and I are gonna get married. Yup, she's got me. Probably end of August next year we reckon, don't we Kim? We'll make sure it's before the vendange. You'll come, won't you? Course, we'll be drinking some of that Clos d'Yvigne stuff. Maybe even some of the sticky gunge too.'

The birds have eaten all the pure noble rot and left only the swollen grapes. I stride through row after row of vines in fury and despair, disbelieving and defeated.

'It's because they could see them,' mutters Edge. 'There aren't any leaves to hide them.' As we speak, it starts to rain.

*   *   *

Jean de la Verrie's son, Xavier, and his daughter, Véronique, come to the *chai* as I wash down a vat. Yesterday's discovery that my noble rot harvest was over is still fresh and painful. It's late afternoon and I imagine they've come for the *barriques* that I had offered their father earlier in the week. I smile at them and they look at me.

'No. It's Uncle Geoffroy,' they say, heads down.

I look at them, uncomprehending.

'We thought you should know. You were his best friend.'

My head has a band of steel around it. It's pressing hard against my forehead and temples and I can't hear what they are saying to me for the high, vibrating sound in my ears.

I have come say goodbye to Geoffroy. As I enter the courtyard of the château, the silence is overpowering, its solemnity and stillness a haunting lament. Madame de la Verrie comes forward and kisses me. '*Pauvre Patricia,*' she says quietly.

Geoffroy's sweet face, his body lying in repose before the funeral in the chateau . . . Madame de la Verrie's face is ashen as she stands next to me, looking at me intently and crying silently as I gaze at him. Hot tears of sadness and regret and anger fall from my eyes at the injustice of it. That he should have gone so quickly without letting us know, when we all loved him so. Monsieur de la Verrie's face is vacant and empty, a witness to the brutality of Geoffroy's sudden death, a heart attack on the train. The fact that he died alone is unbearable.

The bells toll in peals of three.

The church is full on the day of his funeral. And the rain stops. Michel has mown around the church, fastidiously, as Geoffroy would have loved, and flowers are planted at the foot of the trees. I think of him, sitting on his mower day in and day out, perfecting the grounds of the château and his house. Always productive, loving the sun and peace and order, permanently tanned and handsome and always laughing. And loving people.

Most of my *vendangeurs* come to the service, as do all the members of his family. The priest talks of Geoffroy's love of life, his love of Gageac and his family and friends. His eldest nephew and godson, Hervé, Jean's son, reads the lesson, his voice breaking with emotion and love. And we say goodbye to him.

'Bye, Boss,' says Edge as he hugs me. 'Really tough about the birds and the rain. But you have got some barrels of it.' He adds softly 'And Geoffroy's gone to a better place.' Then continues: 'And we'll be back soon. Maybe for a honeymoon?'

He gives me another hug and leaps onto the train as the whistle blows.

We had a frantic dash to the station at Ste Foy la Grande to catch the connection for the TGV fast train to Paris, then Belgium. He and Kim are going to spend two days with his friends who live and work in Brussels and have a baby son, Edge's godson.

Kim and Edge are burdened with the usual enormous backpacks, plus cases of wine and sundry extras, including his Doc Marten boots, which are slung over his shoulder. I'm really sorry to see them go. They can't stay for the *fête de vendange* as they have run out of time. I wave goodbye with regret and a deep sadness that pervades me.

'I'll email you!' he shouts as the train leaves the station. 'And I'll be back!' as his hand waves from the open train window.

The rain continues to fall after the bird catastrophe and although I hang on for another three weeks in the faint hope of another pick, it isn't to be. In a way, it is a relief to finally know. Bruno and Claudie, at least, have succeeded in picking most of their grapes in the short window of opportunity we all had around Edge's birthday, and I did have some barrels of my own from the first two picks before the rain, if not nearly as many as I had hoped.

And somehow, it doesn't matter. Geoffroy's death has rendered superficial all my worries of harvesting, accentuating as it did the fragility of our existence.

Chapter 13

'WHY HAVEN'T YOU GOT A WEBSITE?'

I've only just come to terms with emails and invoices. Edge asked me the same question once. 'I could have a bash for you if you like? Can't be difficult. And I like the idea of being a master, even if its only of webs,' he said.

'It's a good marketing tool,' says one of my bondholders, 'and you could even sell wine on it.' He's particularly enthusiastic about websites and believes the future lies in information technology, and dotcoms and websites in particular. 'It won't cost you much to have a basic site. I'll work on it and let you know what I think. I know someone who's pretty good at creating them and he isn't too expensive.'

I've also got to have a full-time worker. I've worried frantically about whether or not I'll be able to afford one, but now I have decided I can't afford not to. And I can't keep on driving the tractor, making the wine and piling up cartons outside the front of the house, as well as filling barrels, racking, cutting off shoots and all the rest, to say nothing of the ever increasing amount of paperwork required for every aspect of my life now. Each time I begin to understand how to fill in forms and declarations, the format changes or the forms are abandoned completely and something entirely different replaces them and I have to start over again. Even with the help of Jean down at La Ferriere who fills in countless forms for me, and of Juliana and Michel who pile cartons onto lorries, I'm swamped.

'*Oui*, you need a good, solid worker,' says Gilles. 'But he must be reliable. I'll vet him. And don't forget, 8 o'clock start means in the vines at 8 o'clock, not just at your front door!' '*Milledieu*,' he adds, ' I know how hard it is to find someone dependable.'

I have two false starts. '*Milledieu*, he's useless!' says Gilles of the first. 'He wouldn't have lasted the morning *chez moi*!' Even allowing for Gilles's exacting requirements, I have to agree that my new worker is, without a shadow of a doubt, useless. Working up and down the rows together, he has to stop at the top of every other one to roll and light up a cigarette. At 9 am, he wanders off to his car to find a flask and pour himself a coffee. And at 2.30 pm, he takes an afternoon break.

The second is not much better, and goes the same way as the first.

'*Ça va*,' says Gilles of the third and current worker. 'Much more like it. Still, he's not as fast as you or me,' he continues. 'But at least he seems to know what he's doing.'

A contract is signed and Monsieur Alain Beylat becomes my full-time worker.

Ma Patronne
Great that you're on line at last. Brussels was good. Back to the daily grind. May be out soon for a quick weekend after the rugby in Paris. Here's one of my poems:

> *Blue Suede Shoes*
> *Today*
> *I look like Elvis*
> *Except with a fridge*
> *Errr. Fringe*
> *And no quiff*

Edge

The *Guide Hachette* have written. Would I like to taste at the next judging for local wines? And would I judge at the second tier tasting in Paris for the national awards? I decide I must be on someone's list.

The first *Guide Hachette* tasting is, like the first of any of the tastings, I now know, intimidating and I listen carefully as we are instructed on the requirements. The wines we are tasting have passed an initial tasting to eliminate those that don't meet the quality desired. Our judging will be stringent, in panels of three. Our comments must be full and our marks must reflect the quality of the wines. If any wine is outstanding and all three members of the jury agree it merits a *Coup de Coeur*, the forms should be kept separately and presented to the officials at the end of the tasting.

My panel is tasting Monbazillacs and we have fifteen bottles in front of us. Silence descends as everyone concentrates.

The first wine is simply extraordinary and I recognise it instantly as Bruno and Claudie's. It's mellow, golden colour and clearly defined bouquet are pure. With honey and apricots and butter, it's soft and voluptuous. The harmonious balance of fruit and acidity, alcohol and sugar is a sure sign of its purity. It fills my mouth with flavours, complex yet subtle and its richness, never cloying, but powerful, lingers in my mouth long after I have spat out, as the fruit flavours and aromas persist.

I wonder what we're supposed to do if we recognise someone's wine, then decide it can't possibly be a bar to its obvious merit as a great wine. I don't hesitate in marking it at five, the maximum, and placing a cross next to the *Coup de Coeur* box. Those that follow are all of high quality, but none match the excellence of the first.

At the end of the tasting, our panel confers and I am delighted when I see that the other two members are in agreement about the first wine.

The wines submitted for a *Coup de Coeur* are re-judged by the grand jury and as I leave, I see them gathering, with a group of people outside the tasting room, most of them the better wine-makers, awaiting the eventual results.

My reds are ready for bottling. Jean Marc and I taste the *assem-blages* of the wines in the *chai*. I've decided to make two reds; one from my original vines and the other from Gilles's. One of my bondholders has suggested giving all the wines names. I choose Petit Prince for the red from my vines, after Saint Exupéry's book, and Rouge et Noir, the title of a book by Stendhal, for Gilles's.

'*Tres littéraire,*' laughs Jean Marc. 'And French writers too.'

The classic dry white is Princesse de Cleves, after a story by Madame de La Fayette, and the barrelled white is Cuvee Nicholas, after Nick Ryman. 'If you're going to make a barrelled white, take my advice and don't over oak. Be discreet,' he said. I did take his advice and he was right.

Bel Ami is chosen as the name for the rosé, after Maupassant's book, the jaunty, rakish character in the story reflecting the char-acter of the wine.

'*C'est une affaire qui marche,*' pronounces Monsieur Capponi, arriving to bottle the reds and looking around the *chai*. With a stab of pain, I think instantly of Geoffroy, who used to say that all the time. How I miss him. I look down the road to where the cross sits. Thanks to him and the land donated by the de la Verrie family, the commune has widened it. I look onwards to the cypress trees in the cemetery, sitting like sentinels, guarding and protecting his grave. He loved them as much as I do.

*   *   *

We are bottling all the red; two days of intense activity with pallet after pallet sitting on the lawn next to the church, with the perennial problem of where to house them. Once again, they are crammed into every available space. Even with the *dépôt de stockage* and the atelier, there's still not enough space. We transport the overflow to Gilles's house and deposit it in his *chai*. I look at it all. How am I going to sell this much?

In fact, my stress levels rise each time I catch sight of them — when I open the door of the atelier, when I look into the *dépôt de stockage*. It is a nagging worry that I'd managed until now to put to the back of my mind. The wine is a reality; I've got to sell it.

I go to the telephone to start marketing, to the post office for packages to send samples off, to the computer to send letters to as many contacts as I can find. I need more exposure to the market. I need more customers.

Applying myself to marketing is another steep learning curve. My initial foray into the market was so discreet it was barely perceptible to anyone, least of all my potential buyers. I soon found that selling something properly requires an energy and drive quite different from that needed to grow, nurture, pick, vinify and bottle. I begin by looking up names and telephone numbers of wine merchants in England, then ring them up, suggesting I send samples. It is time consuming and takes concentration — and is hard work.

Spring is here again. Suddenly the landscape is green and verdant, with fresh young leaves unfolding on the trees. Daffodils are in bloom and the spring sunlight spreads its harmony and fresh beauty everywhere. The dogs leap around the field, basking in the regeneration that comes with the season. With it come the birds with their song. With it also come the baby snails working their way steadily up the vines to the fresh young leaves.

'*Quoqillooo! Patricia!*' Juliana shouts, waving from her bedroom window. *'Qu'il fait beau! Salut mes enfants!'* at Sam and Luke who are looking up at her from the bottom of the garden, tails wagging in expectation of titbits. She has flung open all her shutters and duvets and mats are hanging from the upstairs windows again. 'Chloe! Chloe! Be quiet! I'm coming!' as Chloe barks at her. Spring is definitely here.

Alain, my worker, is about to mow the grass in the vines and the man who I buy my weedkilling products from has arrived to discuss this year's programme of *traitements*. Pierre Charlot is also here, having dropped off some wine for a *dégustation*.

And Gilles has just appeared. *'Salut! Comment ça va?'* Yes, he'd like a coffee and *'Oui, je vois que tu as commencé passer le girobroyeur.* Make sure he works down the rows three times over at my house too, and I hope he's in the vines at 8 o'clock every morning? And I notice he's missed a *morceau* in the middle of that row too!' he says, pointing at the row just worked.

'I've come to see you about *ma mère's* house' he continues, when everyone has gone. 'You know you have first charge on it in the contract?' I do know and indeed, it would be out of the question for anybody else to have it, surrounded as it is by my vines. Anyone, that is, apart from Gilles.

Doesn't he want it, I ask? Doesn't he miss Gageac?

'Oui, un peu, I've spent my life here. But our new house is right next door to work for Pamela, and I've got everything nearby. *Non,* I don't want to live back there.' He's quite settled in his new life and new home, he says, and the wall he has been building is almost finished. It's being heightened to stop some of the noise of the tractors which pass day and night during the vendange, delivering grapes to the cave co-operative. *'Ooui,* day and night!' he shouts, smiling. 'I can't get away from the vendange! *Milledieu!'* then laughs.

'The rest of the time, it's quiet and without problems,' he

continues. 'We need to discuss a price.' Then he adds, 'but it'll be correct, you know that.' And I do know that.

'*C'est la Galerie Lafayette ici!*' pronounces Juliana, laughing and bringing with her a delicious concoction of vegetables and Parmesan. '*Bonjour mes enfants!*' she shouts to the dogs. 'How is Gilles? He's walking much better now.'

We both agree how much we miss Madame Cholet passing by on her way to the cemetery, bucket on her arm with her jaunty walk and saying '*Bonjour Madame*', always with some comment on the vines, or life or Gilles and his stones.

'*Ce n'est pas pareil,*' it's not the same without her, says Juliana.

Or without sweet Geoffroy and his visits back and forth from his house to mine.

'*Mais!*' she exclaims. '*Nous*, we are alive and *la vie est belle*, isn't it?' She is laughing.

Yes, it is.

A message on the answerphone. It's from the Japanese who had come to taste last year. 'Madame Atkinson? Monsieur Komata. Je voudrais soixante Petit Prince, soixante Rouge et Noir, soixante Princesse de Clèves, soixante Cuvée Nicholas, cinquante Saussignac.' He has left his telephone number in Tokyo.

'*Oui* , the boat is leaving next week, Madame,' he says when I ring him. 'Can you have the wine ready?

'Will you need an *acquis* for the twenty-five cases?' I ask, never having sent wine to Japan and not knowing what system they use.

'Non, non, Madame, we would like 290 cases. Can we have your bank details so we can transfer the money today?'

Working the vines on my own, *épamprée*, in one of the *parcelles* to the right of Gilles's house, I straighten up for a moment and am once again amazed by the beauty before me. Gentle sun and silence save for birdsong, with a sweeping view of the valley in front of me. The rows of vines are green bands of soft leaf and form, stretching into the distance, up and down the undulating slopes. To the right and in the far distance is an orchard of fruit trees, dazzling white and pink. Further still is Bergerac, with other small villages to the left, culminating in Gardonne. Ahead and slightly to the right is the hill of Rouillac, Monsieur de Madaillan's château hidden behind a copse of trees. Fresh greens and gentle forms everywhere, the whole caressed by a spring haze.

I think fleetingly of Monsieur de Madaillan as I continue the *empamprage*. He, too, has died, like Madame Cholet and dear Geoffroy. Pierre, Geoffroy's father, passed by the house yesterday on his walk along the road surrounding Eida's field. As he walks with difficulty now, two nurses assist him. And yet, at ninety-six, he still goes on, now having outlived his son Geoffroy and his small daughter, Suzanne. He has spent his entire life here. I wonder what he has seen – the carriages that brought his mother up the hill to the château she grew to love so much, and the great changes that are now part of Gageac's history. Even I have a history here now, intertwined with the de la Verrie family and the Cholets and the vines and the land in front of me.

A fax arrives. James Nicholson Wine Merchants, Co Down. *Tasted the samples, which we liked very much. Fifty cases of Princesse de Clèves, fifty of Petit Prince, fifty cases of Cuvée Nicholas.* That's Ireland to add to Japan, as well as Britain.

'Patricia?' It's one of my bondholders on the phone. 'Gave a bottle of your Rouge et Noir to one of my chums in Switzerland. He's a wine importer and he loves it. He's sending you an order by email.'

I open my emails to find his order. There is also an email from Edge.

Subject: Lance Armstrong

Hello

Sorry for not writing to everyone individually but I don't have so much time on line at the moment as I need to stay in hospital for a while so they can complete the tests they're running on me to get a suitable course of treatment.

It seems overwhelmingly likely that the mass in my right lung is a tumour and the hope is that it is going to be lymphoma which is extremely treatable and I've had a doctor say that in that case it would be ninety-nine to a hundred percent likely to clear up.

I didn't want any Chinese whispers getting round that I had lung cancer, or that I had an extra arm growing out of my left ear. Mostly I'm feeling fine and I've got lots of cakes and books to read. I'm learning lots and hoping to go on to win the Tour de France in 2002.

It's not nice news to have to tell but I'm confident that I only have one thing to do at the moment and that as soon as we know exactly how I've got to do it, then we'll be getting on with it. It's only multi-tasking that I have any trouble with and now that I don't have to do the marathon, I'll just get on with getting myself fixed up in time for the summer party season.

Don't be sad as I fully intend to live.

Love from Edge

Subject: Lance Armstrong

Dear All

I'm having to adjust to the news that I got yesterday that not only confirmed it was cancer in my lung and liver, but that it was an adenocarcinoma, which is a bad one. I'm hoping to get an appointment ASAP at the Marsden to find out exactly what course of treatments I have open to me. There must be one somewhere that gives me a hope.

At the moment I'm feeling pretty down and I'm sorry for having to share the news with you, but having the truth of it out in the open helps me a lot. Thanks for your words of support so far and I know that it is difficult for you to think positively, but all I ask is that you just think positive things for me.

Don't rush out a reply to me if you don't know what to say, just send me a *positive thought and I'll build up a war-chest so that I can take this bastard on.*

Take care
Love from Edge

Subject: Maillot Jaune
Dear All

Delete this email if it's stuff you don't want to hear. The cancer has spread to my liver and is reasonably well advanced and this had led to an initially unfavourable prognosis. I'm being transferred to the Marsden. I will shortly be looking for the best doctor there that I can get my hands on. I have taken in that the prognosis is somewhat bleak and am looking for something to keep me alive for as long as possible. At the moment, it seems that an outright cure is not on the cards, so I need to just keep myself around until there is one.

With this in mind, the situation is relatively simple and I am currently researching information and putting together a Dream Team from whom I demand unconditional support in whatever I want, with no guarantee of anything in return.

Incredibly, I've started to fill the posts already and have been overwhelmed by the offers of support from everyone. What I need is more relevant information about my condition and opinions about treatment and hospitals and diets — all things will be carefully considered. If anyone knows a colour therapist, a witch doctor or an expert on leeches, there are still places available on what I hope will be a winning Dream Team.

Sorry for sending such a rambling long email but this is important to me. I'm just a little bit scared, I'm feeling very positive and anxious to get syringe-fulls of noxious chemicals into my veins and start the fight. There cannot be a negative outcome of this. And if you're wondering what a tumour looks like, it's a shrivelling insignificant lump of flesh that's being dismantled by white blood cells and love — of life.

The last few days have been probably the most amazing of my life — filled with overwhelming support and love and I would like to pass on the feeling I have gained; that of knowing how much I love life and want to live. It's unstoppable, unrelenting, there are dips and fears but I have never felt more alive.

It's going to be a fantastic party at the end. We'll have carrot and celery juice with macrobiotic rice canapés . . .

I'm just sending a bit of the love that I've been sent over the last few days. And saving the rest so that I can save my precious life.

Will send updates periodically.

This has been a message of love from a Man on the Edge

(I always knew I'd been given the name for a reason.)

Edge's cancer spread to his brain. His Dream Team gathered momentum and soon numbered one hundred and thirty people, all of whom loved Edge and supplied him with information on alternative medicines, diets and treatments, as well as support and love and the will to live.

Incidentally, he writes, *if anyone knows anyone who might want to advertise on these updates, we're experiencing week-on-week growth of five to ten new Dream Team members, in a dynamic ABC1 social group with exciting e-business possibilities.*

The Dream Team members were only allowed to send positive messages. *'Yellow cards to those who foul up!'*

I'm sorry about the last one,' I say in one of his phone calls to me. 'Can't I have another colour?'

'There's only red after that, Boss,' he says.

His overwhelming mood of hope, his lack of fear, his courage and his sense of humour were remarkable. The Dream Team received email after email.

When he begins his chemotherapy course: *It's ironic as I'm suffering no side effects at the moment and feel I could start a little gentle training for next year's Tour de France.*

Later on when the effects have run their course and towards the end: *Sat down to look at a book with Kim and absent-mindedly pulled my fringe out. It had to happen some time and now it leaves me with several options for the wedding day on June 16th; Mr Sheen, a large top hat or a tasteful wig .*

*   *   *

'Hi Patronne.' Edge is on the phone; a quiet, serene voice. He has just had a blood transfusion, he is saying, and is feeling a sense of calm. He has called to say he'd like to come out for the vendange. 'Guess I won't be doing much. But we can put Kim to work. You are coming to the wedding, aren't you? I'd really like you to be there.'

Edge died three days later, hours before his wedding. Peacefully, with Kim, his parents and Becky at his side.

Chapter 14

'GRANDMA!' AMY RUSHES INTO MY ARMS. CHANTAL, AMY AND BETH are here, with Becky.

'Where's Sam?' Beth is now talking. 'Sam!' she exclaims as Luke bounds over to her.

There is a cry of *'Quoqilloo mes enfants!'* and Juliana is here too. *'Oh, la la! Qu'elles sont belles!'* as she kisses Amy and Beth. She has brought sweets, two giant bags full. Chantal notices them with dismay *'Je sais!'* Juliana adds quickly. 'That's why I'm giving them to you to dole out,' and hands them over, pointing at the packets to the children and laughing. 'For you!' she says, in English.

'The garden looks lovely, Mum,' says Chantal.

We are at the back of the house, in Sam and Luke's domain. The trees have grown. Gone are the poppies and wild flowers. In their place are trees and shrubs and a courtyard like the front of the house. The hydrangeas are in full bloom, as is the orange blossom and the roses. The honeysuckle emits its heady perfume, in competition with the privet, which is also in blossom.

'Can I water them, Grandma?' asks Amy.

Becky, Chantal and I plant a white rose for Edge in the garden. Amy and Beth lend a hand, piling soil around its base. 'He's gone to heaven, Grandma.' says Amy, 'Hasn't he, Becky?' looking up at her.

'He certainly has,' I reply, as Becky can't.

*   *   *

The *salle de dégustation* is now finished and definitely gives a better impression to customers, looking much more professional than the kitchen, where previously tastings were done. And visits from people wanting to buy wine have increased radically too. Amy and Beth join them and listen as I explain the varieties, the tastes and the prices.

'I can smell oranges, Grandma!' exclaims Amy, sniffing the Saussignac 'and honey!'

The courtyard in front of the tasting room is also emitting perfumes. Pepper from the tamarisks, delicate rose scents from the rose bushes planted by Michel and the overpowering perfume of privet in flower. The wisteria has finished its first spectacular flowering, but now has a second one, more discreet, but still scented.

'*Mais, alors, elles ont grandi!*' they've grown, exclaims Odile as she steps into the garden and kisses the children. She, too, has brought them a packet of sweets, but this time a more manageable quantity and size. '*Et si on descend prendre un pizza ce soir?*'

We do.

It is a hot, sultry evening and tables are arranged outside, behind the pizzeria. I recount to the children Geoffroy's story of the angel with châteaux in her apron as we finish off with ice creams, replete with sparkling decorating sticks, most of which are now carefully positioned in the girls' hair.

Gageac looks resplendent. The château's grounds are mown to perfection and the church has been cleaned and decked with flowers. Outside, flowers have been planted at the base of the trees and the pathway has been strewn with lily petals and laurel leaves. Jean de la Verrie's daughter, Virginie, is to be married here, the first de la Verrie to do so for one hundred years.

It is hot, very hot. A canopy has been hurriedly erected to the right of the church doorway as protection from the sun. Guests arrive. Women wearing elegant hats and colourful clothes stroll up the path to the church with their escorts. Jean de la Verrie arrives, looking mondain and handsome in a morning suit, as do Xavier and Hervé, his two sons. Elizabeth de la Verrie, wearing a large but delicate hat and looking chic, is helped out of the car, followed by their two daughters, looking exquisite and happy.

Monsieur and Madame de la Verrie, *le Comte* and *la Comtesse* arrive and walk slowly up the path and into the cool interior of the church. I think of them and their constant kindness and generosity and of how sorrowful the last event we attended in the church was. Geoffroy would have loved to be here today.

Smiling and dazzling, the bride walks up the aisle. The church is bursting with people, who have spilled out onto the lawn and under the canopy outside. Once the ceremony is over, the bell rings joyously as the bride and groom walk out of the church into the bright sunshine, and a hubbub of noise and laughter and gaiety.

We go back to Jean de la Verrie's house, Jean Brun. It looks as magical as the château. Rose petals and laurel leaves are strewn along the paths here too, and huge bouquets of flowers line the pathway. Long tables, covered with crisp linen cloths and filled with delicacies are set outside for the *vin d'honneur*, to which the whole village has been invited.

The feast continues until five in the morning, with champagne and dancing and laughter.

We are planting an extra two hectares of vines. One *parcelle* will be behind Gilles's house, the other at the edge of some existing vines. A bulldozer has broken up the bedrock that sat fifty centimetres beneath the soil, our tractors have worked and worked the land, and I now understand Gilles's obsession with stones. *'Oooui, il faut passer le bulldozer!* You can't do it otherwise!' he had shouted when I had announced that we would simply work the land with the tractor and plough. 'You'll break the plough and exhaust the tractor! It'll be impossible for the young plants to survive! A complete waste of money! *Et ça côute cher, en plus!'* I took his advice and booked the bulldozer.

The stones add up to a veritable mountain. They are horrendously large, some of them measuring two metres in diameter. They are lifted out of the land and placed at the edge of the *parcelles.*

'Impressionant, n'est-ce pas?' Gilles remarks as we look at the pile. 'Told you, didn't I?' He continues: 'And that's not all. You now have to gather up the stones that are going to appear each time you work the land!'

He is right. We heave stone after stone into trailer after trailer. It takes five of us three weeks to do it, and each time we drag the plough through the ground again, more appear. And each time we remove them, the soil that was sitting above them slips into the holes. Eventually, I decide to pass a fine plough over the two *parcelles* because otherwise we will never be finished heaving stones into trailers and will eventually have no soil left anyway, to say nothing of wearing out our backs, knees, elbows and necks, comme Gilles.

'Ce n'est pas gênant,' it's not a problem, remarks the planter when he comes to inspect the land, which in spite of our best efforts, still has stones scattered all over it. It is just as well, as 2900 young plants were ordered much earlier in the year, and they have been growing and awaiting plantation.

The planters are here with them now, a lorry load of exquisitely beautiful young plants.

'Grandma! Come and see!' shouts Amy.

They are marking out rows in the sauvignon *parcelle* with string, narrow straight rows, aided by a machine that guides the string and signals whether or not the markings are correct and aligned.

Two workers are carefully placing plants at one-metre intervals along the rows and behind them a machine is boring holes for them to be put into. A team of eight men, working under the broiling sun on the last official day for planting, the 31st of July, is methodically and diligently progressing along one row after another, occasionally stopping to take a swig from their water bottles or to wipe their brows.

There is no shade. Just an expanse of barren land with soil, the scattered white stones reflecting the heat upwards. We watch them in fascination, working up and down rows. I wonder why they deposit the young plants along the row before the hole has been dug, then decide it must be so that the worker manipulating the boring machine knows where to dig. Another worker follows him, gently deposits the plant in the hole and then waters it.

'You'll need to irrigate them again in two days,' says the planter when they are finished. 'I'll come along next week to have a look at them. You must water underneath the leaves, otherwise you'll burn them.'

I look at them more closely. The leaves are only two inches or so above ground level. Sitting under the burning sun that beats down on them relentlessly, they are minute, delicate and beautiful. I marvel at their resilience, at the beginning of yet another cycle of life; a hope for the future.

It is hot, very hot, as I leave the *parcelle* and wander up towards the dilapidated cross and Madame Cholet's house on my way back home. In the distance is Gilles's house, about to be restored. I look out again from the ridge, deep into the far distance of the fertile valley. It has permanence and beauty, enduring through the changes the country has seen in its tumultuous history; through the lives of ordinary people whose roots in the land have touched its history. I think of Madame Cholet with her deep understanding of the land, and of Geoffroy with his desire for order and his love of life. And of Edge with his friendship, his loyalty and courage.

In front of me are rows and rows of vines. Graceful, sinuous curves, architectural in form, their large leaves lush and green, protecting and nourishing, soaking up the sun and sweetening the harvest. The sun shines, hot and relentless, onto the grapes. They hang like jewels, perfectly crafted. Each vine represents the march of time, the past, the present and the future.

I feel Edge's presence beside me suddenly, his smiling face, his hope, which never waned. 'Investigate your options, Boss,' he had said. 'You've got loads of them. There's nothing that can't be achieved, you've just gotta stay positive. And go forward.'